Environmental Ecology, Biodiversity and Climate Change

Environmental Ecology, Biodiversity and Climate Change

Towards Sustainable Development

(Core Text for Civil Services Prelims Examination – Paper I)

H.M. Saxena

RAWAT PUBLICATIONS

Jaipur • New Delhi • Bangalore • Guwahati • Kolkata

ISBN 978-81-316-0689-6 (HB)
ISBN 978-81-316-0690-2 (PB)

Reprinted, 2017

Published by
Prem Rawat for **Rawat Publications**
Satyam Apts, Sector 3, Jawahar Nagar, Jaipur 302 004 (India)
Phone: 0141 265 1748 / 7006 Fax: 0141 265 1748
E-mail: info@rawatbooks.com
Website: www.rawatbooks.com

New Delhi Office
4858/24, Ansari Road, Daryaganj, New Delhi 110 002
Phone: 011 2326 3290

Also at *Bangalore, Guwahati* and *Kolkata*

Typeset by Rawat Computers, Jaipur
Printed at Nice Printing Press, New Delhi

Contents

Preface

Environmental issues are central to the political, social, economic, and moral challenges of the 21st century, therefore, a complete knowledge about these issues is necessary for administrators as well as for the common man, not only for their proper understanding but also for the planning and sustainable development.

The UPSC has designed a new syllabus for the Civil Services, both for the pre as well as main examination. A significant component of this new syllabus is the introduction of environmental issues and sustainable development. The topics included are '*General Issues on Environmental Ecology, Bio-diversity, Climate Change*' and '*Sustainable Development*'.

The book, '*Environmental Ecology, Biodiversity and Climate Change: Towards Sustainable Development*', is an attempt to provide relevant and updated literature on the above mentioned topics. The book has been designed in such a way that it will develop a full understanding of these core topics and will definitely help in solving the multiple-choice questions.

Chapter I deals with general issues related to environmental ecology. In this chapter, a brief analysis of environment, ecology, their components, functioning, world ecosystem and problems of environmental degradation have been given. Chapter II is related with 'Biodiversity'. Biodiversity is the foundation of life on the earth, which includes all the species of living organisms. A detailed discussion on all the aspects of biodiversity has been done with special reference to India.

Chapter III is on 'Climate Change', which is the single biggest environmental and humanitarian crisis of our time. The aspects discussed in this chapter are climate effects, prevention measures and international efforts done in this field.

Sustainable Development is the theme of Chapter IV. The present world needs a development strategy that meets the uses of present generation without compromising the ability of future generation to achieve own requirements. The concept of sustainable development and efforts done towards it as well as other related aspects have been discussed.

Based on reliable data and information, and written in simple logical and convincing style, I am confident that the book will be of immense help to the candidates appearing in UPSC Civil Service and other competitive examinations. The book will also be a useful guide for all those who are interested in environmental issues and sustainable development.

I am thankful to the entire publication team of the Rawat Publication for their inspiration, persuasion and support in publication of this book. My thanks to my wife Nirmala Saxena for her moral support. I am also thankful to all the members of my family and friends for their good wishes.

H.M. Saxena

1 Environmental Ecology

General Issues

The term 'environmental ecology' is a combination of two well-known terms, 'environment' and 'ecology', which are different but interrelated to each other. Together, these two determine the existence of all the living creatures on the earth and are also responsible for the growth and development of civilizations right from its dawn to the present day. They will also direct the future course of development.

Environment

The term 'environment' etymologically means surroundings. Literally, it is an English word formed by two words, i.e. 'environ' and 'ment' which means 'encircle' or 'all round'. Thus, environment is a complex of many variables which surrounds man as well as all living organisms. The place where an organism lives is called 'habitat'. It is also known as 'milieu' which means 'total set of

> **Box 1.1**
> **Definitions of Environment**
>
> The sum total of all conditions, agencies and influences which affect the development, growth, life and death of an organism, species or race *(The Universal Encyclopaedia)*.
>
> The entire range of external influences acting on an organism, both the physical and biological (in other organisms) forces of nature surrounding an individual *(Encyclopaedia Britannica)*.

surroundings'. In fact, habitat is a part of the earth's surface where life originates and develops according to the geographical conditions.

The basic components of environment and their characteristics are as follows:

Box 1.2
Components of Environment

1. *Atmosphere:* The atmosphere is a composition of gases, dust particles and water vapours that surrounds the earth. It fulfils the following functions:
   - It sustains life on earth.
   - It saves us from the hostile environment of outer space.
   - It absorbs most of the cosmic rays from outer space and a major portion of the electromagnetic radiation from the sun.
   - It transmits to the earth only ultraviolet, visible, near-infrared radiation (300 to 2,500 nm) and radio waves. The atmosphere is composed of nitrogen, oxygen, argon, carbon dioxide and trace gases.

2. *Hydrosphere:* Hydrosphere comprises of all types of water reservoirs – oceans, seas, lakes, rivers, streams, polar icecaps, glaciers and groundwater. It has the following characteristics:
   - About 72 per cent of the earth's surface is covered by hydrosphere.
   - Out of this, oceans cover 97 per cent area.
   - About 2 per cent of the water resources are locked in the polar icecaps and glaciers.
   - Only about 1 per cent is available as fresh surface water, rivers, lakes, streams and groundwater, fit to be used for human consumption and other uses.

3. *Lithosphere:* Lithosphere is the outer surface of earth. It consists of minerals occurring in the earth's crust and the soil, e.g. minerals, organic matter, air and water.

4. *Biosphere:* Biosphere comprises of living organisms and their interactions with the environment – atmosphere, hydrosphere and lithosphere.

The three types of environment are:

(i) Physical environment
(ii) Biotic environment
(iii) Social or cultural environment

In fact, environment is constituted by the interacting systems of physical, biological and cultural elements, interrelated in various ways, individually as well as collectively. It is a complex phenomenon of several factors and elements which are the product of the forces and processes of nature. The environmental elements are classified as under:

Box 1.3
Environmental Elements

1. *Physical Elements:* Physical elements include space, landforms, water bodies, climate, soils, rocks, and minerals. They determine the variable character of the human habitat, its opportunities as well as limitations. Physical environment refers to the geographical climate and weather or physical conditions wherein an individual lives. The human races are greatly influenced by the climate. Some examples are as under:
   - In the cold countries, such as European countries, North America, Siberian Asia and North-East Asia, the people are of white colour. Likewise, in South Asian and African countries (hot countries), the people are of dark complexion.
   - The physique of an individual depends on climatic conditions as the individual tries to adjust in his/her physical environment.
   - The humans' working efficiency also depends on the climatic conditions.
2. *Biological Elements:* Various elements – plants, animals, microorganisms and humans – constitute the biosphere. Although, physical and social environments are common to the individual in a specific situation yet, every individual has his/her own psychological environment, in which he/she lives. Lewin has used the term 'life space' for explaining psychological environment. The psychological environment enables us to understand the personality of an individual. Both the person and his/her goal form psychological environment. If a

person is unable to overcome the barriers, he/she can either get frustrated or may change their goal for a new psychological environment. By adopting this mechanism, the individual is helped in his/her adjustment to the environment.

3. *Cultural and Social Elements:* Cultural elements – economic, social and political – are essentially man-made features, which make cultural milieu. Social environment includes an individual's social, economic and political conditions in which he/she lives. The moral, cultural and emotional forces influence the life and nature of an individual's behaviour. Society may be classified into two categories as under:
   - An open society, which is very conducive to individual development.
   - A closed society, which is not very supportive to development.

Ecology

Ecology is the science of the relations of all organisms to their environment. The two components of nature, organisms and their environment are not only complex and dynamic but also interdependent. In 1859, St. Hilaire Isodore Geoffory, a French zoologist, proposed the term 'ethology' for 'the study of the relations of organisms within the family and society in the aggregate and in the community'. Similarly, a British naturalist, St. George Jackson Mivart, used the term 'hexicology', which he defined in 1894 as

Box 1.4
Definitions of Ecology

Ecology is defined as the study of the relations of organisms, or groups of organisms to their environment, or the science of the interrelations between living organisms and their environments (Odum).

Ecology is the science of all the relations of all organisms to their environments (Taylor).

Ecology is the science which treats the organisms in relation to their environment (Moore).

Ecology is the science dealing with the relationship of organisms to one another and to other factors that comprise their environment *(Encyclopaedia Britannica)*.

'devoted to the study of the relations between the organisms and their environment as regards the nature of the locality ...'. The term 'ecology' was coined by combining two Greek words, viz. 'oikos' (meaning 'house' or 'dwelling place') and 'logos' (meaning 'the study of') to denote the relationships between the organisms and their environment.

Now ecology is a systematic and well-developed science having several subdivisions and specialized fields as shown in Figure 1.1.

Figure 1.1 *Main Subdivisions and Specialized Fields of Ecology*

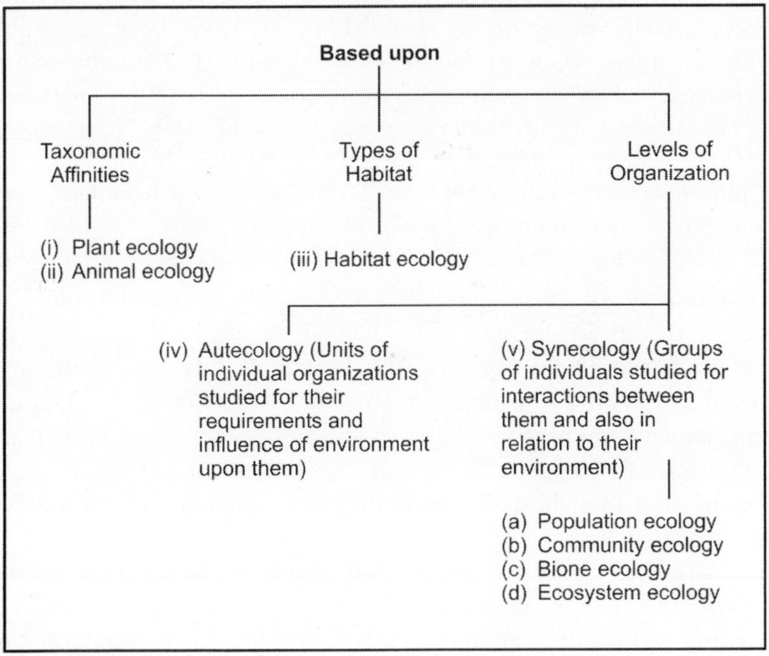

Ecosystems

The whole earth and its parts, viz. land surface, water bodies, atmosphere as well as living organisms are working in an integrated system, which not only directs but also determines the pattern of life and is also an indicator of environmental relationship with living organisms. Each and every part of the earth's surface is having its own ecosystem but at the same time it is also a part of a larger system, which may cover the whole of the continent, parts of various continents or even whole of the earth. The single giant ecosystem is the 'biosphere'. The term 'ecosystem' was coined by A.G. Tansley in 1935, who defined it as 'the system resulting from the integration of all the living and non-living factors of the environment'. He further stated that 'the whole system, includes not only the organism complex but also the whole complex of physical factors forming what we call the environment of the biome – the habitat factors in the widest sense. It is the systems so formed which are basic units of nature on the face of the earth'.

Although the concept of ecosystem is much older than the term itself, many terms parallel to it have been used by bio-scientists, such as *biocoenosis* (Mobius, 1877), *microcosm* (Forbes, 1887), *bolocen* (Friederichs, 1930), *biosystem* (Thienemann, 1939), *geobiocoenosis* (Sukachev, 1944), *ecosphere* (Cole, 1958), etc. However, the term 'ecosystem' is now accepted and most preferred, where 'eco' implies the environment, and 'system' means interdependent, interacting complex.

According to Monkhouse and Small, ecosystem is 'an organic community of plants and animals viewed within its physical environment or habitat'. The habitat or physical environment controls the whole world of organic community, viz. plants and animals. With the change in physical environment, the type of plants and animals too have changed. Similar views have been expressed by Haggett: 'Ecosystems are ecological systems in which plants and animals are linked to their environment through a series of feedback loops'.

In fact, our earth has a giant ecosystem in the form of the biosphere, where abiotic and biotic components are constantly acting and reacting upon each other, bringing forth structural and functional changes in it. The biosphere is a vast ecosystem and difficult to study as a whole, therefore, for the convenience of study, it can be divided into the following major subdivisions:

(i) *Terrestrial*: as forest, grassland, desert, mountains, etc.
(ii) Aquatic (water – open): (a) *fresh water*, which may be *lotic* (running water like spring, stream, rivers) or *lentic* (standing water like lake, pond, swamp, pools, etc.); and (b) *marine*, such as deep sea or shallow sea, delta, etc.
(iii) *Artificial*: (man-engineered) ecosystems.

Components of the Ecosystem

The two major components of the ecosystem are:

- The biotic components comprising of all living organisms.
- The abiotic components including the physical (non-living) environment.

Biotic Components

In an ecosystem, organisms are usually divided into two categories on the basis of their nutritional (trophic) relationship. These are: (a) autotrophic components, and (b) heterotrophic components.

(a) *Autotrophs* are those organisms which can produce their own food. These are green plants (with chlorophyl) and certain bacteria, which obtain their energy from the sun and through photosynthesis produce food. Since these organisms produce food for all the other organisms, they are known as *producers*.

(b) *Heterotrophs* are those organisms which depend directly or indirectly upon the autotrophs for their food, thus known as *consumers*. The consumers are of two types:

   (i) *Macroconsumers* also known as herbivores, carnivores or omnivores. Herbivores are the plant eaters, while carnivores are animal eaters. On the other hand, omnivores eat all kinds of food. They all are phagotrophs which include chiefly animals that ingest other organic and particulate organic matter. These are also known as primary consumers.

   (ii) *Microconsumers* are also known as decomposers or osmotrophs. They are decomposers because of their role in decomposition of dead organic matter. They secrete digestive enzymes to break down food into simpler substances and then absorb the digested food.

Functionally, the biotic components can be divided into three types, viz. producers, consumers and decomposers. Producers produce their own food like plants, which are primary producers, while man and animals obtain food from plants. Similarly, consumers are also primary, secondary or tertiary. Man, cow, goat, etc. are primary consumers, while meat eaters are secondary consumers. Tertiary consumers are those animals which eat meat such as snakes eat frogs, peacocks eat snakes, etc. Among decomposers, those bacteria, fungi, etc. have been included which are responsible for the decomposition of a product and release of inorganic nutrients into the environment.

Abiotic Components

The physical environment with its several interacting variables constitutes the abiotic component of ecosystem. It consists of (a) the lithosphere, i.e. the solid mineral matter of the earth, (b) the water in the oceans, rivers, lakes, etc. or in other words, the hydrosphere, (c) the gaseous mixture in the air (the atmosphere), and (d) the radiant solar energy.

Ecosystem Functioning

The functional pattern of an ecosystem is a very important and basic aspect of study, because all its components are dynamic in one way or the other and are responsible for the creation of a unique state of man-environment relationship as well as a habitat, different from other regions. Ecosystem functioning can be studied under three heads: (i) energy flow, (ii) productivity, and (iii) biogeochemical cycling.

Energy Flow in an Ecosystem

Energy is defined as the ability to do work and is the essence of life. Without energy transfers, there could be no life and no ecological systems. The energy use for all the life processes is derived from solar radiant energy. The behaviour of energy is described by the laws of thermodynamics: (a) energy may be transformed from one type into another but is never created or destroyed, and (b) no process involving an energy transformation will spontaneously occur unless there is a degradation of the energy from a concentrated form into a dispersed form, etc.

The energy reaching the earth's surface is transformed and/or absorbed by plants and other organisms. It is used by the green plants during photosynthesis by converting the light energy to chemical energy and making it available to other organisms as food. Such energy transfers along the food chain were studied for the first time by Lindeman in 1942. He stated that all the functions of an ecosystem could be explained in terms of energy by the knowledge of two attributes of each trophic level – the level of energy storage and the efficiency of energy transfer.

Productivity of Ecosystem

The productivity of an ecosystem refers to the rate of production, i.e. the amount of organic matter accumulated in any unit of time. It is of primary, secondary and net productivity.

(a) *Primary Productivity* is defined as 'the rate at which radiant energy is stored by photosynthetic and chemosynthetic activity of producer organisms (chiefly green plants) in the form of organic substances which can be used as food materials'. Primary productivity is of two types:
   (i) *Gross Primary Productivity* is the total rate of photosynthesis including organic matter used up in respiration during a particular period.
   (ii) *Net Primary Productivity* is the rate of storage of organic matter in plant tissues in excess of respiratory utilization by plants during the measurement period.

(b) *Secondary Productivity* is the rate of energy storage at consumer level. It actually remains mobile from one organism to another and does not live *in situ* like primary productivity.

(c) *Net Productivity* refers to the rate of storage of organic matter not used by the consumers. Thus, it is the rate of increase of biomass of the primary producers which has been left out by the consumers.

Food Chains in Ecosystems

The transfer of food energy from the producers, through a series of organisms (herbivores to carnivores to decomposers) with repeated eating and being eaten, is known as a *food chain*. Producers utilize the radiant energy of sun which is transformed to chemical form, ATP during photosynthesis. Thus, green plants occupy, in any food chain, the first trophic (nutritional) level – the producers level – and are called the *primary producers*. The energy, as stored in food matter manufactured by green plants, is then utilized by the plant eaters – the herbivores, which constitute the second trophic level – the primary consumers level, and are called the *primary consumers* (herbivores). Herbivores, in turn, are eaten by the carnivores, which constitute the third trophic level – the secondary consumers level – and are called the *secondary consumers* (carnivores). These, in turn, may be eaten still by other carnivores at tertiary consumers level, i.e. by the *tertiary consumers* (carnivores). Some organisms are omnivores that eat producers as well as the carnivores at their lower level in the food chain. Such organisms, thus, may occupy more than one trophic level in the chain. This simple chain of eating and being eaten away is known as a *food chain*. A food chain in grassland ecosystem starts with grasses and herbs and goes through grasshoppers, the frogs, the snake, the hawk in an orderly sequential arrangement based on the food habits, whereas in a pond the order would start with phytoplanktons, going through water fleas, smaller fish, bigger fish, birds, larger animal and so on.

Food Webs: Interlocking Pattern of Organisms

However, food chains in natural conditions never operate as isolated sequences, but are interconnected with each other forming some sort of interlocking pattern, which is referred to as a *food web*. Under natural conditions, the arrangement of food chain hardly occurs and these remain indeed interconnected with each other through different types of organisms at different trophic levels. For example, in grazing food chain of a grassland, in the absence of a rabbit, grass may also be eaten by a mouse. The mouse in turn may be eaten directly by a hawk or by a snake first which is then eaten by a hawk. Thus, in nature, there are found alternatives, which all together constitute some sort of interlocking pattern, known as the *food web*.

Figure 1.2 *Food Web in a Grassland Ecosystem*

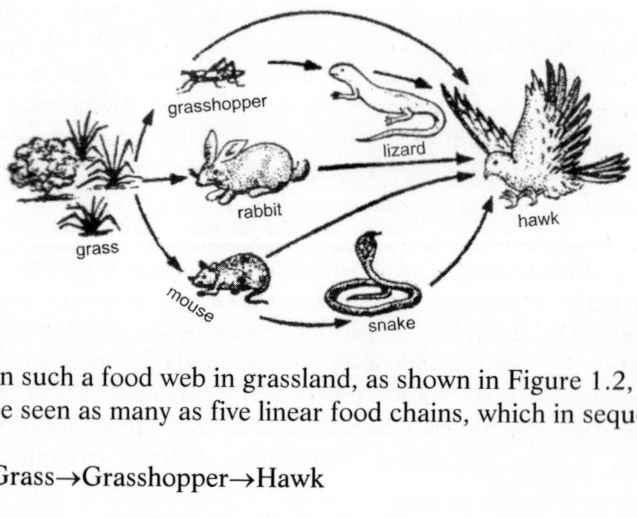

In such a food web in grassland, as shown in Figure 1.2, there may be seen as many as five linear food chains, which in sequences are:

(a) Grass→Grasshopper→Hawk

(b) Grass→Grasshopper→Lizard→Hawk
(c) Grass→Rabbit→Hawk (or vulture or fox or even man, if present)
(d) Grass→Mouse→Hawk
(e) Grass→Mouse→Snake→Hawk

The complexity of any food web depends upon the diversity of organisms in the system. It would accordingly depend upon two main points:

(i) Length of the food chain: Diversity in the organisms based upon their food habits would determine the length of food chain. More diverse the organisms in food habits, more long would be the food chain.
(ii) Alternatives at different points of consumers in the chain: More the alternatives, more would be the interlocking pattern.

In deep oceans, seas, etc. where we find a variety of organisms, the food webs are more complex.

Biogeochemical Cycles

The chemical elements, including all the essential elements of protoplasm, tend to circulate in the biosphere in circular paths from environment to organisms and back to the environment. These move in a more or less circular path, thus are known as *biogeochemical cycles*. The growth of life process requires about 40 elements. Hydrogen, carbon and oxygen are the basic elements, while nitrogen, phosphorus, potassium, calcium, sulphur, magnesium, and iron are the other important elements. All the elements continue to cycle at the ecosystem level; they are also an integral part of the larger and/or global cycle. The cycling of various elements can be grouped into two categories, viz. (i) *gaseous cycles* of elements like hydrogen, carbon, oxygen, nitrogen, sulphur; and (ii) *sedimentary cycles* of those elements which are present in the soil and organisms. The following biogeochemical cycles are functioning in an ecosystem:

(i) Hydrological cycle
(ii) Carbon cycle
(iii) Oxygen cycle
(iv) Nitrogen cycle
(v) Phosphorous cycle
(vi) Sulphur cycle

Ecological Pyramids (Graphic Representation of Trophic Structure)

Trophic structure, i.e. the interaction of food chain, size and the metabolism relationship between the linearly arranged various biotic components of an ecosystem is characteristic of each type of ecosystem. The trophic structure and function at successive trophic levels, i.e. producers, herbivores and carnivores can be shown by ecological pyramids where the first or producer level constitutes the base of the pyramid and successive levels, the tiers making the apex. The three types of ecological pyramids are:

(i) *Pyramid of numbers:* They show the relationship between producers, herbivores and carnivores at successive trophic levels in terms of their members. The pyramid of numbers in a grassland ecosystem is shown in Figure 1.3. In a grassland ecosystem, the producers, mainly grasses, are maximum in number. This number then shows a decrease towards the apex, as the primary consumers like rabbits, mice, etc. are lesser in number than grasses, the secondary consumers, snakes and lizards are even lesser in number than rabbits and mice. Finally, the top consumers (tertiary) like hawks or other birds are the least in number. Thus, this pyramid becomes upright. The shape of the pyramid of numbers varies in forest ecosystem and parasitic food chain.

(ii) *Pyramid of biomass:* The pyramid of biomass in a grassland ecosystem is upright (Figure 1.4), while it is inverted in a pond

Figure 1.3 *Pyramid of Numbers*

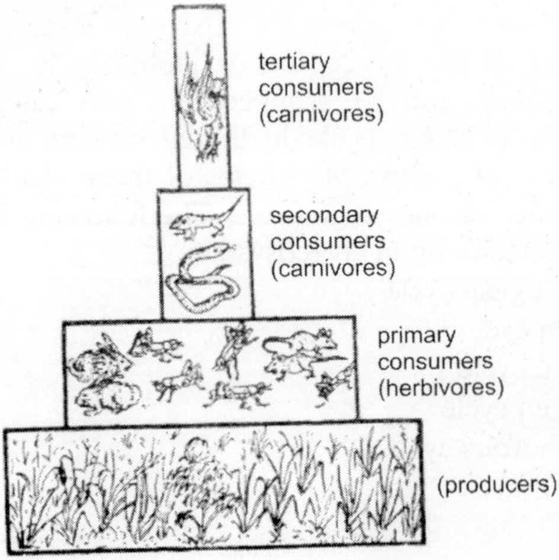

ecosystem, where producers are small organisms; their biomass is the least, and this value gradually shows an increase towards the apex of the pyramid, thus making the pyramid inverted in shape (Figure 1.5).

Figure 1.4 *Pyramid of Biomass in a Grassland Ecosystem*

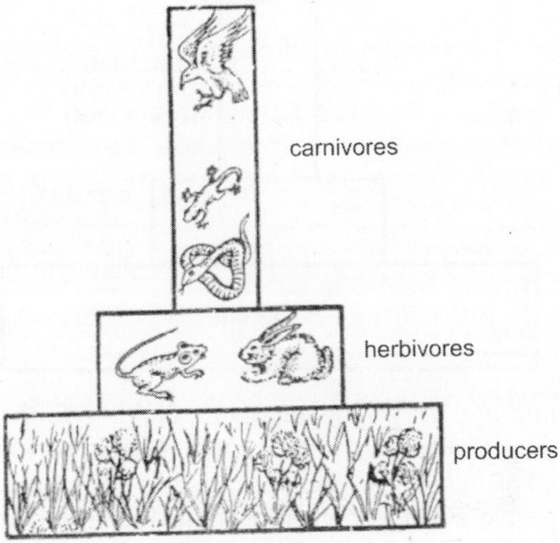

Figure 1.5 *Pyramid of Biomass in a Pond Ecosystem*

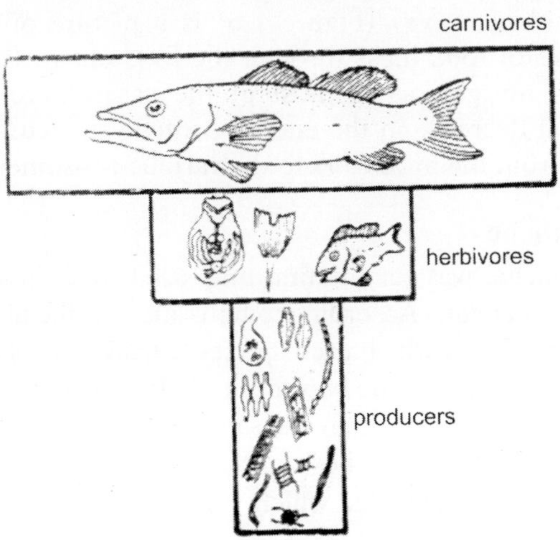

Figure 1.6 *Pyramid of Energy in a Ecosystem*

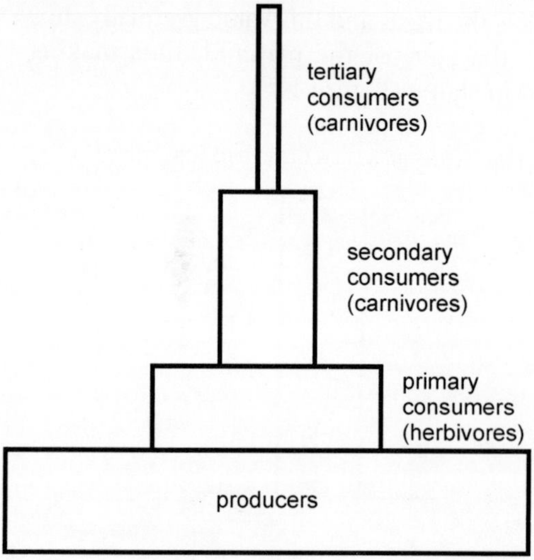

(iii) *Pyramid of energy:* The energy pyramids give the best picture of overall nature of the ecosystem. Here, number and weight of organisms at any level depends not on the amount of fixed energy present at any one time in the level just below but rather on the rate at which food is being produced. In contrast with the pyramid of numbers and biomass, which are pictures of the standing situations (organisms present at any moment), the pyramid of energy (Figure 1.6) is a picture of the rates of passage of food mass through the food chain. In shape, it is always upright, as in most of the cases there is always a gradual decrease in the energy content at successive trophic levels from the producers to the various consumers.

Ecological Niche

The term 'niche' was for the first time used by Grinnel (1917) to explain microhabitats. According to him, 'niche is the ultimate distributional unit, within which each species is held by its structural and instinctive limitations ... no two species in the same general territory can occupy for long identically the same ecological niche'.

The term 'niche', whose dictionary meaning is 'a place, employment, or activity for which a person is best fitted' should never be confused with habitat. The habitat is the place where an organism lives, or the place where one would go to find it. The ecological niche on the other hand, is a more inclusive term that involves not only the physical space occupied by an organism, but also its functional role in the community (i.e. trophic position occupied), and its position in environmental gradients including other conditions of existence. These three aspects of ecological niche are generally designated as: (i) the spatial or habitat niche (physical space occupied), (ii) the trophic niche (functional role, i.e. trophic position), and (iii) the multidimensional or hypervolume niche (position in the environmental gradients). Thus, ecological niche of any organism concerns not only the space where it lives, but also what it does (i.e. how it transforms energy, behaves, responds to and modifies its physical and biotic environment), and how it is constrained by the other species. The concept of ecological niche is of much significance in ecology in terms of differences between species at same physical place, or at different places, or the same species at more than one location. The species concept is thus mainly considered.

Biomes: World Ecosystem Types

Biomes are the regional-scale assemblages of ecosystem on the surface of the earth and occupied by distinctive plant and animal communities. The pattern of biomes is essentially controlled by primary production, which provides the energy base for the biological communities that make each biome. There are wide variations in productivity, and thus variation in the types of plants. These are controlled by a series of biotic environmental factors. The world pattern of terrestrial biomes is closely related to the global climate.

Box 1.5
Definitions of Biome

A mixed community of plants and animals (a biotic community) occupying a major geographical area on a continental scale (Goudie, 1994).

A major regional ecological community of plants and animals extending over a large area, with a uniform climate and vegetation (Simmons, 1979).

Box 1.6
Terms Related to Ecology/Ecosystem

Ecological Balance: The achievement of a study state by an ecosystem, as in a climax community. This theoretical concept is rarely accomplished in practice as controlling forces, such as climate and soil, rarely remain constant.

Ecological Crisis: A state of human-induced ecological disorder that could lead to the destruction of ecological conditions.

Ecological Efficiency: The ability of the organism at one trophic level to convert to their own use the trophic level directly beneath them.

Ecological Succession: It is a process of community development in ecological time, from seasons to centuries. It is often described as 'plant succession'.

Ecological Footprint: The area of land functionally required to support a human society.

Ecological Imbalance: The destabilization and destruction of a frangible environment, often as a result of economic development.

Ecological Relationships: The relationships between organisms in an ecosystem. All organisms in an ecosystem are connected. There are various ecological relationships in which some are oppositional and some are symbiotic. Peridation and competition are oppositional relationships; while mutualism, commensalism, amensalism and parasitism, etc. are symbiotic relationships.

Carbon Dating: It is a process of finding out how old something is by analysing the amount of carbon dioxide present in it which was decayed.

Biochore: It is the geographical area, supporting a distinct plant or animal life.

Climatic Climax: Ecological succession is an orderly event and depends entirely on environment. When ecosystems reach their climax known as climatic climax then after that species structures will be stable. But other group of ecologists do not believe in such orderly system and status that climax is determined by a host of factors. Such mosaic of climax is known as polyclimatic climax.

Major Biomes of the World

The global pattern of biomes is varied due to the differences in physiography, climate, natural vegetation, soil and water bodies. Table 1.1 lists the global pattern of biomes in the world.

Table 1.1 *Major Biomes of the Earth*

Biome	Mean Annual Temperature (°C)	Mean Annual Precipitation (cm)	Producers	Typical Primary Consumers	Net Production g.m2/year	Other Characteristics
Aquatic/Marine biome						
Lotic (flowing waters)			Rooted aquatic plants	Invertebrates, fish	250	Well aerated
Lentic (lakes, ponds)			Floating plants	Invertebrates, fish, phytoplankton	250	Poorly aerated
Swamps (wetland with trees)			Rooted plants, trees and graminoids	Invertebrates, fish, some mammals	<2000	High productivity
Marshes			Herbaceous (graminoids) wetlands	Invertebrates, fish, specialized rodents	<2000	—
Estuaries			Rooted plants, phytoplankton	Invertebrates, fish	<1500	High productivity
Mangrove swamp			Mangroves (various species)	Invertebrates, fish	1100	Littoral or estuarine
Littoral (shoreline)			Phytoplankton	Invertebrates	—	Moderate productivity
Neritic (continental shelf)			Phytoplankton	Invertebrates, fish	—	Moderate productivity
Reefs			Algae symbiotic with corals	Fish, corals and other invertebrates	<2500	Tropical, high species richness, high productivity

Cont'd...

...Cont'd

Pelagic (open ocean)			Phytoplankton	Invertebrates, fish	125	Low productivity
Terrestrial biome						
Temperate forest	–3–20	6–250	Deciduous trees, shrubs, annual and perennial herbs	Browsers, frugivores		Highly seasonal, cold winters, hot summers
Temperate grassland	–5–20	50–130	Grasses	Grazers, seed eaters	600	Fire-, drought-, wind-influenced, extremes, burrows
Tropical savanna and scrub	17–30	50–150	Scattered trees over grassland	Grazers, browsers	900	Fire prone
Taiga (boreal coniferous forest)	–5–5	5–200	Needle-leaved trees	Browsers, seed eaters	800	Minimal under story, high litter mass
Tundra	–15–5	10–100	Sedges, grasses	Grazers (lemmings, musk ox)	140	Permafrost, many breeding shorebirds
Chaparral	–5–20	50–75	Hard-leafed scrub	Browsers, seed and fruit predators	700	Summer drought, winter rain, fire prone
Desert	–5–30	0–50	Stem- or leaf-succulents	Browsers, seed predators	90	Water-rich tissues protected by spines, etc.
Tropical rainforest	17–30	25–450	Tall buttressed trees, vines, epiphytes	Specialized herbivores, trees, animals	2200	Little seasonality

General Issues Related to Environmental Ecology

In the beginning of the 21st century, environmental and ecological issues have emerged as a major global concern for human survival. In the opnion of the World Commission on Environment and Development, the future is to face an ever-increasing environmental decay, poverty, hardship and an evermore polluted world.

Now it has been realized that environmental problems are not limited to local, regional and national level but there are several global environmental problems. The scientific and technological revolution has given multiple facilities to mankind but at the same time it is responsible for depletion of resources, excessive use of fossil fuels, deforestation, desertification, loss of the fertility of soil, changes in atmospheric conditions resulting into serious problems like greenhouse effect, depletion of ozone layer, rise in world temperature, etc. All these and many other problems have a global dimension and the world community has already realized this fact that only cooperative efforts can at least, minimize these environmental problems if not solve it. A brief review of the general issues related to environmental ecology is given here:

Deforestation

Forests are not merely a natural resource to supplement multiple human requirements but are a very vital aspect of the environmental system. The forest plays a pivotal role in balancing the ecosystem. Growing industrialization, urbanization and ruthless exploitation of forests have created chaotic conditions and severe phyto-geographical and environmental imbalances. The use of forests has been linked with mankind from times immemorial basically for food, shelter and clothing, but population exodus and stresses on forest resources created a condition where natural growth of forests and their maintenance is incapable of coping with the demand.

Tropical forests are the richest forest resources of the world, their distribution has been shown in Figure 1.7. But rapid deforestation of these forests have become an environmental issue. Scientists estimate that one-fifth of the world's tropical rainforest was destroyed between 1960 and 1990. They claim that 50 years ago

Figure 1.7 *Global Distribution of Tropical Forests*

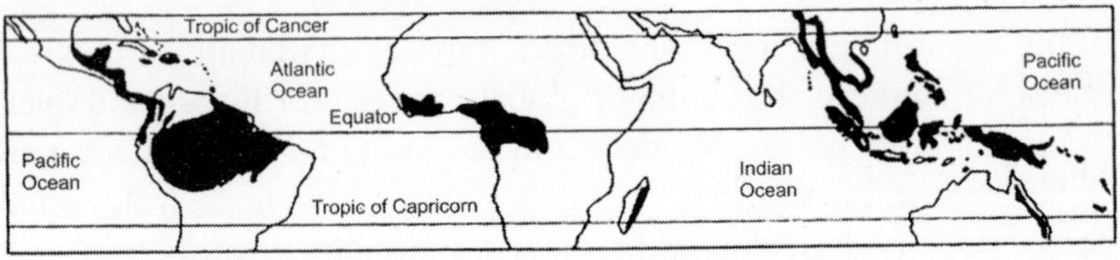

rainforests covered 14 per cent of the world's land surface but now only cover 5 to 7 per cent and that all tropical forests will be gone by the middle of the 21st century. In 2005, UN Food and Agriculture Organization (FAO) estimated that the total forest area is decreasing at about 13 million hectares per year. The rate of deforestation is much higher in countries like Brazil, Cameroon, Costa Rica, India, Indonesia, Myanmar, Philippines, Thailand and Vietnam. Since pre-agricultural times, the world's forest have declined by one-fifth, from 5 to 4 billion hectares. Temperate forests have lost the highest percentage of their area (32–35%), followed by sub-tropical woody savannahs and deciduous forests (24–25%), and old growth tropical forests (15–20%). Tropical evergreen forests are now under the highest pressure. Although most of the tropical forests are located in inaccessible areas and sparsely populated zones, now these forests are facing the tragedy of destruction.

Tropical deforestation has three direct causes, which often act together in the same area. The first cause is permanent conversion of forest to agricultural land. Logging is another cause, although deforestation depends on the methods used and on what follows the logging. A third cause of deforestation is the demand for fuel wood, fodder, and other forest products where the resources cannot meet the demands. These direct causes as well as grazing, forest fire and drought are often interrelated and are exacerbated by government's economic policies, population pressure and poverty. The main cause of deforestation is the commercial exploitation of forest.

In India, deforestation is going on at a very fast rate. The story of overexploitation of forests started from World War II, when the British started clearing forests to meet their needs. In the post-independence period, the process of exploitation has been accelerated to meet developmental projects of railways, industries, mining, river valley projects, defence, etc. The urbanization and expansion of agricultural activity has also contributed to the exploitation of forests. The commercial exploitation of forests is still going on in spite of several legal provisions to restrict it. Maximum deforestation has occurred in Madhya Pradesh, Chhattisgarh, Maharashtra, Odisha, Andhra Pradesh, Jammu and Kashmir, Rajasthan, and Himachal Pradesh.

The impact of deforestation on the environment can be seen in the form of:

(a) Increase in temperature,
(b) Lesser precipitation,
(c) Increased rate of soil erosion,
(d) Increase in frequency and volume of floods,
(e) Loss of soil productivity,
(f) Extinction of several species,
(g) Non-availability of several essential forest products, and
(h) Imbalance in ecosystem.

The following methods should be taken for the conservation of forests:

(a) A proper ratio between forest cutting and plantation of trees should be maintained;
(b) Actual growth of trees be supervised;
(c) Strict prohibition of the cutting of trees for fuel be imposed;
(d) Alternative media for wood such as moulded plastic, etc. be developed for furniture and other domestic articles;
(e) Protection from forest fires,
(f) Development of national parks and game sanctuaries;
(g) Social forestry be developed;
(h) Plantation of trees according to local ecological conditions;
(i) Forest survey be conducted and classification be done accordingly; and
(j) Long-term master plan for forest conservation and development be prepared.

In fact, forests can be protected through public movements like 'Chipko'.

Desertification

Desertification is a complex process affecting more than six hundred million people and approximately one-third of the earth's land surface. Desertification, to an ecologist or environmental scientist, is a systematic phenomenon involving deterioration of the ecosystem due to change in the pressure of human adaptation or use system. It is a process through which fertile or non-desertic land is converted into dry, infertile or desertic land.

According to Mann (1977), 'desertification can be defined as a package of processes which brings about certain basic changes in a desertic terrain. This involves an interplay of climatic, adaphic and biotic factors which may (a) work through time starting with a climatic accident and ultimately resulting in the formation of a desert, or (b) lead to accentuation of the arid conditions within an already desertic terrain, or (c) cause a creep of the desertic conditions within a desertic tract towards its adjoining areas'.

The UN Conference on Desertification (UNCOD) has defined desertification as 'the diminution or destruction of the biologic potential of the land, which can ultimately lead to desert-like conditions'. There is no environmental problem in the world that affects poor people as extensively as land degradation or desertification, which affects more than 6.1 billion hectares, about 35 per cent of the earth's land area (Figure 1.8).

Asia, with 1.3 billion hectares of degraded dry lands, suffers the most in case of desertification. But in terms of severity, Africa and North America are even worse off, because nearly three-quarters of

Figure 1.8 *Threat from Desertification*

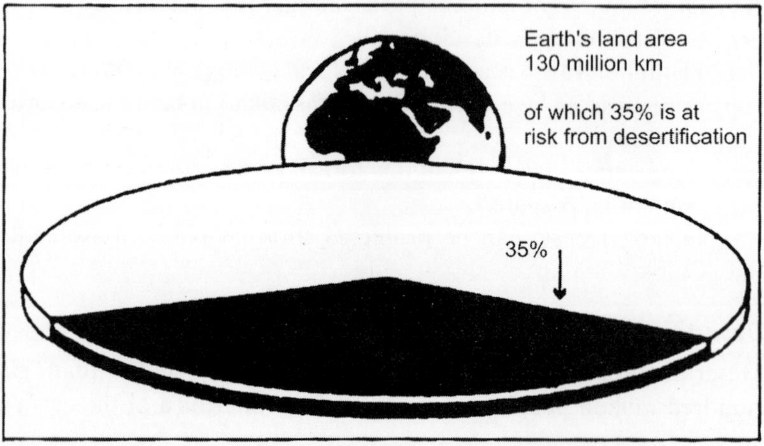

dry lands are affected. In fact, desertification is a major environmental problem and has worsened by the persistent droughts reported in the northern hemisphere during 1988. In Africa, drought and desertification rank as the foremost environmental concerns; heavy rains broke the drought of the previous year with catastrophic floods and an unprecedented desert locust plague. During the last two decades, there have been at least 10 million environment refugees from the dry lands. Desertification is a matter of global dimension.

In India, the arid zone covers about 12 per cent of the country's geographical area and occupies over 320,000 sq km of hot desert located in parts of Rajasthan (62%), Gujarat (19%), Punjab and Haryana (9%), and Andhra Pradesh and Karnataka (10%). In these regions as well as in other arid regions of the world, due to the high pressure on grazing lands and overexploitation of resources, degradation of environment is growing day by day. Increased desertification problems are also evident in the form of salinity hazard, seepage and waterlogging problems in irrigated areas of the desert. The increased pressure of livestock results in overexploitation of resources leading to soil erosion and removal of surface vegetation. The monitoring of deserts is necessary and should be done by remote sensing methods, both orbital and aerial. In order to control desertification, the following steps should be taken:

(a) Ban on cutting of desert vegetation,
(b) Plantation of ecologically suitable plants,
(c) Development of pasture lands,
(d) Control on overgrazing,
(e) Stabilization of sand dunes,
(f) Proper use of available groundwater resources,
(g) Management of land use and farming practices,
(h) National watershed development programme should be intensified, and
(i) Public awareness should be generated.

Soil Erosion

Soil is usually defined as 'any part of earth's crust in which plants root' and is formed as a result of long-term process of complex interactions between climatic, topographic and biological factors. Bare rock surfaces are exposed to various types of physical, chemical and biological processes which lead to physical and chemical disruption. Physical processes of weathering include action of water, temperature, glaciers and gravity which cause weathering of rocks through such processes as wetting, drying, heating, cooling, freezing, glaciation, solution and sand blast, etc. The chemical processes of weathering include hydration, hydrolysis, oxidation-reduction, carbonation, etc. All these processes disintegrate the soil but their impact on the environment is limited. On the other hand, soil erosion on a large scale leads to the degradation of the environment. It may be caused due to natural processes as well as by man also. The main types of soil erosion are as follows:

(a) Normal or natural erosion,
(b) Fluvial erosion,
(c) Aeolian or wind erosion,
(d) Marine erosion, and
(e) Accelerated soil erosion by human actions.

Actually, normal soil erosion is a continuous process and its impact on environment is also limited, but when it occurs on a large scale due to water and wind erosion, it not only degrades the environment but is also responsible for the expansion of wasteland due to gullies and ravines, as has been in case of the Chambal region of Madhya Pradesh and Rajasthan. Four types of soil erosion due to water have been identified, these are:

(i) Sheet erosion,
(ii) Rill erosion,
(iii) Gully erosion, and
(iv) Riparian erosion.

Soil erosion due to wind is a common phenomenon in arid and semi-arid regions and is often a cause of desertification or desert expansion. The process of soil erosion is always accelerated by man either by overgrazing, faulty land practices or by deforestation.

Figure 1.9 shows the areas of soil degradation in the world through deforestation, overgrazing and also due to overexploitation of vegetation.

Figure 1.9 *Soil Degradation: The Hot Spots*

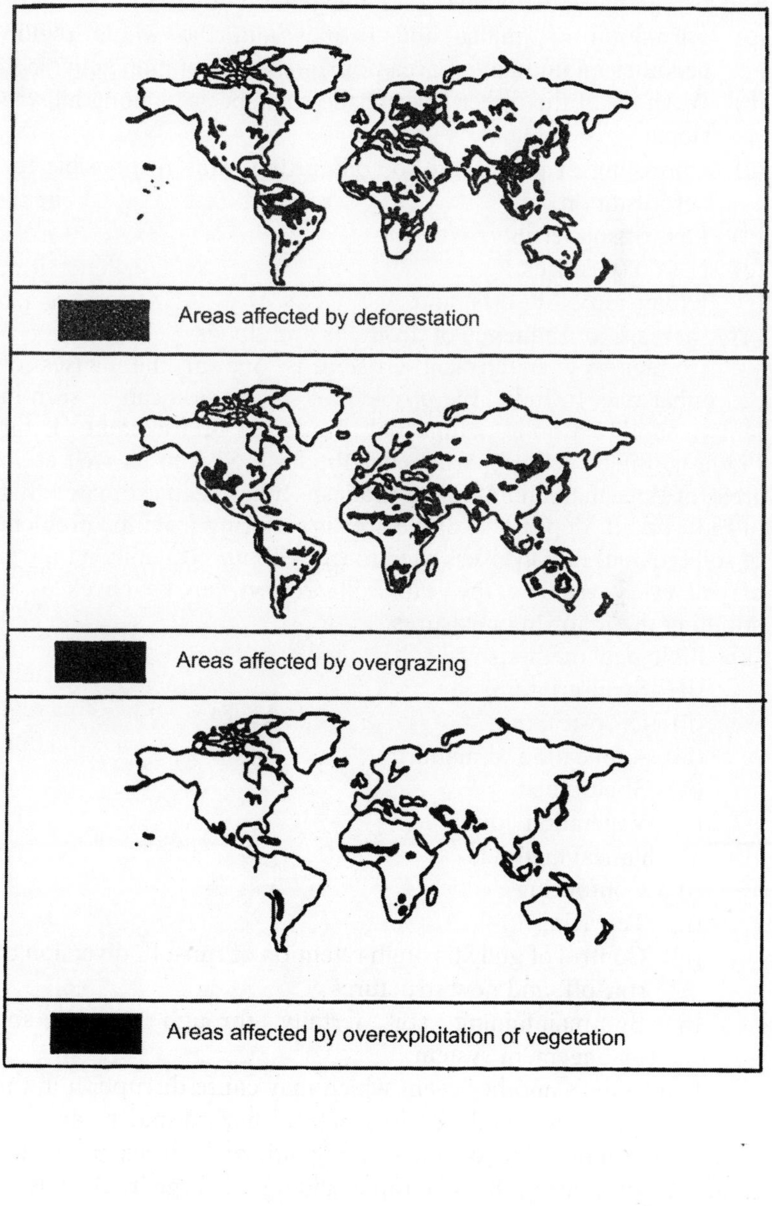

The impact of soil erosion on the environment is both local as well as regional. The following are the general impacts:

(a) Formation of gullies and ravines and the whole region becomes an undulating area not suitable for human activities;
(b) Washout of the upper layer of the soil, specially along hillside slope;
(c) Uprooting of plants due to soil cutting, thus responsible for deforestation;
(d) Loss of soil fertility;
(e) Loss of pastures;
(f) Expansion of deserts; and
(g) Increase in frequency of droughts and floods.

In fact, too much soil erosion is one of the causes of eco-imbalance. In India, the problem of soil erosion can be seen in Jammu and Kashmir, Himachal Pradesh, Uttarakhand, Uttar Pradesh, Bihar, Assam, Arunachal Pradesh, Sikkim as well as in areas of Rajasthan and Madhya Pradesh. It has been estimated that in India about 53 per cent of the area is suffering from the problem of soil erosion. Indian rivers used to carry about 600 million tonnes of soil every year, to the sea. Soil erosion can be checked by adopting the following measures:

(a) Biological methods
    (i) Strip-cropping
    (ii) Crop-rotation
    (iii) Application of manure
    (iv) Shelter belt
    (v) Vegetation cover
(b) Mechanical methods
    (i) Contouring
    (ii) Terracing
    (iii) Control of gully through retention of run-off, diversion of run-off, and new structures
    (iv) By maintaining soil fertility through proper soil management system

Landslide is another event which may cause disruption in the general ecosystem. Landslide is a natural hazard and means any down slope movement of mass of regolith or bedrock under the influence of gravity. It is a rapid sliding of large rock masses beginning their descent as unit of blocks, without internal flowage.

When the landslide is in the form of huge blocks, it is known as 'rock avalanche'. In Switzerland, Norway or Canadian Rockies, villages built on the floors of steep-sided valleys have been destroyed by the sliding of rock masses. In India, reports of rockslide often appear in the news in hilly states like Jammu and Kashmir, Himachal Pradesh, Uttarakhand, and North-East states. The environmental hazard of rockslides is extremely high for towns located on the floors of glacial troughs and at the heads of fiords. The blockage of roads by landslide is a common phenomenon but sometimes it also blocks the course of the small rivers, thus, causing much damage.

Environmental Pollution

With the advancement in science and technology all over the world, environmental pollution has increased many fold and now has become an environmental issue of multidimensional nature. Now pollution is not only a regional problem but has increased to national and international level and is responsible for large scale environmental degradation. The various forms of environmental pollution are air, water, noise, land, and radioactive pollution.

(a) Air Pollution

Air pollution may be defined as an imbalance in the quality of air so as to cause ill effects. According to Maxwell, 'our enormously accelerated abuse of the atmosphere has become a health hazard and a threat to life, damaging both plants and animals in areas polluted with poisonous fumes, dust, and smoke'. The industrial, automotive and domestic activities have resulted in increasingly outrageous insults to the atmosphere. In air pollution, the pollutants include gases, solids and liquid particles of both organic and inorganic chemical classification. The different types of pollutants are continuously introduced into the atmosphere and are removed by natural processes of cleansing. But when pollution exceeds the atmosphere's self-purifying capacity, accumulation of pollutants occurs causing serious hazards for environment and other organisms including humans. The following are the main reasons for the increasing rate of air pollution:

(i) Poisonous gases and other particles emitted from industries without any treatment;

(ii) Heavy increase in the number of automobiles and their emission (smoke and other dangerous gases);
(iii) Increased use of chemicals and petrochemicals;
(iv) Population concentration in cities;
(v) Fast rate of deforestation;
(vi) Tests of experiments of atomic weapons;
(vii) Tests of chemicals and biochemical weapons;
(viii) Space research and satellite wastes; and
(ix) Unorganized mining and traditional practices on the use of fuelwood, etc.

Certain natural events such as volcanic eruption, dust storms, etc. are also a cause of air pollution. In order to understand the nature of air pollution, it is necessary to know the various sources of pollution.

Sources of Air Pollution

Air pollution is the result of the combined effects of several pollutants. These pollutants are associated with each other and also react with other elements, therefore, it is difficult to categorize them, but for proper understanding they can be divided into the following categories on the basis of their origin, nature, size, impact, etc.

1. Particulate matter can be divided into two types, viz. natural and man-made. The natural form of particulate matter is the result of volcanic dust and gases, mineral dust, sea-salt crystals. Another class of natural particulate matter is smoke from forest fires and grass fires. Living plants release pollens and spores into the air, these are organic compounds. From forest trees, certain hydrocarbons called terpenes are also released into the atmosphere.

   Man-made particulate matter comes from many sources, but the major source is from the combustion of hydrocarbon fuels – petroleum products, coal, peat and wood. Combustion of solid wastes is another source. Other kinds of matter introduced into the atmosphere are industrial chemicals, fly ash, refining fossil fuels, mining and smelting ores, as well as pollutants discharged through quarrying, farming activities, etc. Use of various types of solvents and also radioactive elements are the cause of air pollution created by man.

2. Another classification is primary and secondary pollutants. The primary pollutants are those gaseous and other solid microparticles inducted to the atmosphere. These pollutants are emitted and as such are not found in the air. The most important gaseous pollutants are carbon monoxide, oxides of sulphur, hydrogen sulphide, hydrocarbon, oxides of nitrogen, ozone and other oxidants. Secondary pollutants are the result of chemical reactions. Evidences and experiments indicate that exhaust gases of automobiles contribute more in the formation of secondary pollutants. For example, oxides of nitrogen produced in the combustion of petroleum and other fuels emitted into the atmosphere yield ozone in the presence of sunlight. It is to be noted that ozone is not emitted as such into the atmosphere but formed only from primary pollutants.
3. According to chemical composition, air pollutants can be divided into organic pollutants and inorganic pollutants. Others have divided them into solid, liquid and gaseous pollutants. The gaseous pollutants are carbon monoxide, sulphur dioxide, hydrogen sulphide and organic sulphide, hydrogen fluoride, hydrogen chloride, oxides of nitrogen, aldehydes and, organic acids, etc. Particulate pollutants consist of both solid and liquid particles. They vary in size from 0.01 microns to 20 microns. Dust, fume, mist, spray, smoke are included in this category.
4. According to source type, pollutants can be classified as being produced from:
   (i) Combustion,
   (ii) Transportation emissions,
   (iii) Industrial processes,
   (iv) Use of solvents, and
   (v) Radioactivity.

Combustion process yields particulates such as fly ash, smoke, and oxides of sulphur and nitrogen. The amount of sulphur dioxide emitted depends upon the contents in the fuel. High nitrogen yields larger quantities of oxides of nitrogen. Carbon monoxide is also emitted from combustion. The other contaminants that are produced by combustion include acids and aldehydes.

The simplest form of combustion is the use of fuels in domestic use. In India as well as in many developing countries, wood, coal, cowdung and kerosene oil are commonly used. All these materials yields carbon dioxide, carbon monoxide, sulphur dioxide, etc.

Automobiles may be considered as the main source of air pollution, specially in urban areas. Automobile exhausts release smoke and to a certain extent lead particles also. Smoke contains the gaseous pollutants like carbon monoxide, hydrocarbons and oxides of nitrogen.

The rapid rate of industrialization has resulted in more and more air pollution. Various industrial processes release almost all types of pollutants into the air. Some industries like cement, iron and steel, fertilizer, petrochemical, etc. are of great concern because of the difficulty in controlling the emission of pollutants from them. Acid rain has become a great threat to the environment. The use of solvents is increasing with the growing use of paints, spray, polish, etc. Due to the presence of hydrocarbons in these materials, air pollution is caused, which is dangerous for health. Similarly, spray of pesticides in agriculture is also responsible for air pollution even in rural areas. Main sources of air pollution and their impact are indicated in Box 1.7

Nuclear material, when released into the air, is hazardous for all living organisms. Nuclear weapon testings, nuclear reactors, chemical processing plants, research institutes and hospitals contribute many radionuclides to the atmosphere.

Box 1.7
Main Pollutants of Air and their Impact on Health

	Pollutant	*Sources*	*Health Hazards*
1.	Carbon monoxide	Automobiles: due to incomplete combustion/burning of carbon in the fuel of internal combustion engines.	Displaces oxygen in blood, reduces amount of oxygen in blood tissues. Dulls mental performance, cause accident proneness, extra burden on anaemic heart and lung patients.
2.	Particulate (soot, smoke, fly ash, etc.)	Stationary fuel combustion in industrial processes.	Occurrence of photo-chemical smog which can cause widespread deaths in a few days.

	Pollutant	Sources	Health Hazards
3.	Oxides of sulphur	Mainly industrial processes involving burning of coal and oil as fuel.	Temporary or permanent injury to the respiratory system, irritate the upper respiratory tract and lung tissue.
4.	Hydrocarbons	Automobiles, fuel combustion in electric power plants, industrial processes, combustion in solid waste disposal and agricultural burning.	Gaseous hydrocarbons in atmosphere are not toxic but help in forming photo-chemical smog.
5.	Oxides of nitrogen	Burning of fuel at very high temperatures as in transport vehicles, power plants and industrial boilers.	Nitrogen oxides combine with hydrocarbons in the presence of ultraviolet rays of the sun to form secondary pollutants called photo-chemical oxidants.
6.	Others	Photo-chemical oxidants	Photo-chemical oxidants (feroxy-acetyl nitrogen aldehydes and acrolein) cause eye irritation and flu, when they combine with particulates to form smog.

Acid Rain

Acid rain is the outcome of sulphur dioxide gas (SO_2) released into the air by the combustion of fossil fuels. This readily forms sulphuric acid (H_2SO_4) in the minute water films of suspended droplets in the air over cities. The washout of sulphuric acid by precipitation results in rainwater with an abnormally high content of sulphate ion, a condition known as acid rain. Nitric acid, formed by reactions involving the pollutant nitrogen oxides, may also contribute to the acidity of rainwater.

The degree of acidity as measured in the laboratory is 5.65 owing to the dissolution of atmospheric CO_2 in it. Rain, having a pH value of less than 5.65, the value resulting from the equilibrium between atmospheric CO_2 and pure distilled rainwater at 25°C, can be termed acid rain.

In recent years, it has been pointed out by water chemists that pH value of rain in North-western Europe and USA is going down to 3 and 5, which is an indication of acidity in rain water. The pH numbers are on a logarithmic scale, these values mean that rain in these areas is now, often 100 to 1000 times more acid than before. After studying rainwater for 20 years, US scientists have reported average pH value of about 4. They have also observed values as low as 2.1 in rainwater at certain localities. The acid rain is now not only the problem of industrially developed world but is also seen in countries like India.

Some of the adverse environmental effects of acid rain are: excessive leaching of nutrients from plants; foliage and the soil; disturbance of balance of predators and prey in aquatic ecosystems; various metabolic disturbances to organisms; acidification of lakes and streams, and the corrosion of structures.

Smog

'Smog' is the name given for 'smoke fog'. The term was first used by H.A. Des Voeux in 1911 in a report on smoke for deaths. During the autumn of 1909 in Glasgow, Scotland, 1,063 deaths were attributed to smoke and fog. In 1930, in Meuse valley of Belgium, a large number of people were made ill and many were killed by smog. The 'black fog' that settled on London in December 1952 caused 3,500 to 4,000 deaths. It was the result of fog accompanied by thermal inversion over most of the British Isles. In Pennsylvania, smog killed 20 people and about 40 per cent of the population became ill. The residents of the Kanto Plains area of Yokohama, Japan, were afflicted with an accumulation of smog during winters of 1945 and 1946. Similar type of attack was reported in New Orleans, Louisiana, in 1959. In recent years, with the increase in atmospheric pollution, the instances of smog resulting in serious injury or deaths have also become common.

The most irritating and injurious components of smog are the products of reactions in the atmosphere between oxygen, ozone and emission pollutants. The mixture of these products is called 'photochemical smog'. These are the reactions in which oxygen, ozone, nitrogen oxides and hydrocarbons produce those compounds which are toxic and irritating.

The constituents of smog are quite toxic and are responsible for respiratory and cardiac difficulties. Eye irritation is the most common symptom of smog injury. Similarly, bronchial asthma, chronic bronchitis and pulmonary emphysema are associated with smog. It is also injurious to animals and plants and is one of the main causes of environmental degradation.

Measures for Controlling Air Pollution

Some of the measures to control air pollution are as follows:

1. The forest cover should be protected by restricting deforestation and through adoption of afforestation programmes. Trees are the best controllers of air pollution. It is an accepted fact that 33 per cent of the land area should remain under forest/tree cover. It will help in controlling air pollution and also help in maintaining the ecological balance.
2. There must be a 'green belt' around every township and village. Similarly, industrial areas should be surrounded by green belts.
3. The main source of air pollution is the automobiles, therefore, their engines should be redesigned in such a way that their emissions cause minimum pollution. Several steps have been taken and some technology has also been developed, but still it is in a very preliminary stage. Apart from technological change some steps on individual level should be taken by the auto owners, such as:
   - (i) Regular servicing of the vehicle,
   - (ii) Change or replacement of old engines,
   - (iii) Use of filters and after burner,
   - (iv) Engine should be well tuned, and
   - (v) Mixing of anti tibs in diesel and petrol, etc.
4. The use of steam engine by the railways should be stopped or minimized. As far as possible, electrically operated rail engines should be used, which are pollution free.
5. In countries like India, traditional use of fuelwood, coal, etc. should be checked and newly devised smoke free furnaces should be used.
6. The industrial areas should be located at a certain distance from residential areas. The selection of the site should be done with consideration of air direction.
7. Steps should be taken for immediate check on forest fires. For this purpose, proper arrangement and sound infrastructure is necessary.

8. In industries, arrangements for pollution control should be done. Only after full arrangements of effluent treatment is done, permission for production should be given.
9. With the help of chemical reactions, harmful effects of pollution can be controlled. Measures for such methods should be taken and the methodology and material should be available to the concerned persons.
10. Some methods of controlling air pollution are filtering, setting, dissolving, absorption, etc. For these methods, cheap devices should be developed.

Actual prevention of air pollution is not as simple as explained above. However, much of the pollution could be minimized or controlled without undue cost by careful planning. For this, the following steps will be helpful.

Control of Gaseous Pollutants

The equipments which can be used to control gaseous pollutants are classified as combustion, absorption and adsorption equipments. Combustion is applicable to pollutant gases which are oxidizable. In petrochemical, fertilizer, paint and varnish industries, combustion control equipments are useful. Absorption is a diffusional process in which the transfer of gas molecules into a liquid phase takes place. Such equipments include spray chambers, packed towers and sieve plate contractors. Absorption is the means of controlling air pollution which occurs due to some gases and vapours and inflammable compounds which cannot be treated by other means.

Control of Emission from Motor Vehicles

The emission control techniques include tune-ups, catalytic reactors and engines modifications. A high air-fuel ratio will reduce the concentrations of both carbon monoxide and hydrocarbons. The modified engines have an efficient system of burning of the fuel. At present, there is no system of reduction of sulphur dioxide; although, researches in this direction are in progress throughout the world.

Control of Aerosol Emissions

This can be controlled by arresters and scrubbers. Arresters include inertial separators, filters and precipitators. The electrostatic precipitator is considered to be the most effective device for preventing the emission of dust from fuel gases and is the standard equipment for large power stations.

Control by Fuel Selection and Utilization

Coal and oil are the primary fuels in which smoke, grit and sulphur dioxide are the major pollutants. Coal pollutes more because it releases more smoke and carbon into the air. Instead of coal, oil can be used. But in oil, the amount of sulphur dioxide emission is higher. Therefore, fuel selection should be done properly and the harmful impact should be restricted by inducing other chemicals. In order to control smoke, the coal is pulverized before being used. When oil is used for fuel, it is essential that the ratio of air should be maintained constant, in order to prevent smoke emissions.

Control of Air Pollution by Site Selection and Zoning

Selection of an industrial site is the most important factor through which impact of air pollution can be minimized. The industrial site should be selected considering: 1. residential areas, 2. nature of industries, and 3. climatic conditions, specially the direction of winds. The site of each and every industry should have an approval from the department of environment. It should be approved by taking into consideration the long-term perspective rather than immediate effects. Within the city, the location of industrial zone is a critical aspect and town planners should have a check over its location as well as unorganized growth.

Apart from these controls, legal controls are also there. The UNO and WHO have issued guidelines to its members for the formulation of laws regarding prevention of air pollution and accordingly all the countries of the world have enacted certain laws for control. In India, the Prevention of Air and Water Pollution Act, 1974, the Air Prevention and Control of Pollution Act, 1981, and the Environmental Protection Act, 1986, have been enacted for air pollution control. These laws should be imposed properly; apart from this, the most important aspect is public awareness about air pollution because the government alone cannot prevent pollution. It can be checked only through the combined efforts of the government, NGOs and the public.

(b) Water Pollution

Water pollution simply means contamination of water due to any external material, or in other words, introduction of something into natural water which makes it unsuitable for human consumption. WHO has defined water pollution as 'any foreign material either from natural or other sources that may contaminate the water supply and makes it harmful to life, cause of their toxicity, leads to reduction of normal oxygen level of water, causes aesthetically unpalatable effects and spread of epidemic diseases'. Thus, defined it as 'any unreasonable contamination of water which lessens its value to man'. In general, 'water pollution may be defined as the adverse change in composition or condition of the water such that it becomes less suitable for the purposes for which it would be suitable in its natural state, the changes include, physical changes, chemical changes and biological changes'.

Water pollution is caused by several sources, which are not independent in nature but interact with one another. Generally, one or two factors become prominent and may be considered the primary source of water pollution. The various sources of water pollution can be classified as follows:

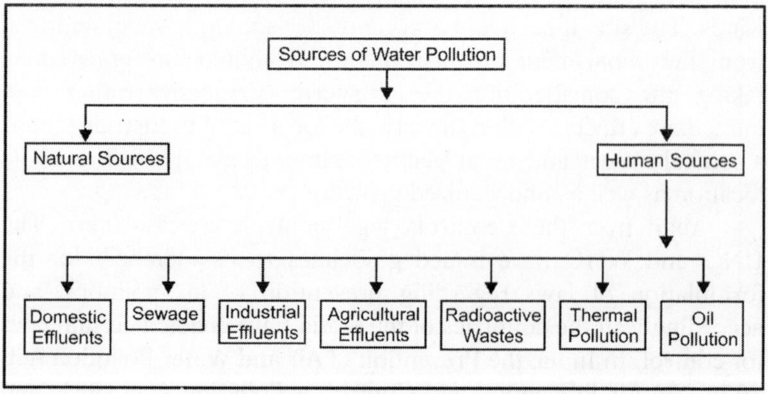

Water pollution contaminants and their impact is listed in Box 1.8.

Measures for Controlling Water Pollution

As described earlier, water pollution now has become a worldwide problem. However, developed countries have initiated several

Box 1.8
Water Pollution Contaminants and their Impact

Source or Type of Contaminant	*Impact*
1. Degradable Wastes	
Domestic and municipal sewage and other oxygen demanding industrial wastes – they are acted upon by the bacteria and oxygen dissolved in water and are reduced to inorganic form, as also in quantity.	Depletion of dissolved oxygen in water is harmful for fish and other aquatic life.
2. Plant Nutrients	
Phosphatic and nitrogenous compounds dissolved in water from industrial wastes, or washing away of fertilizers due to heavy rains from fields.	Stimulate growth of algae/plankton and other plants resulting in clogging of waterways and rendering the water unfit for human use.
3. Infectious Agents	
Polluted, stagnant water breeding grounds for parasites, bacteria, viruses of all kinds.	Cause water-borne diseases and outbreak of epidemics such as ameobiosis, dysentery, cholera, typhoid, etc.
4. Synthetic Organic Compounds	
Such as pesticides, agricultural chemicals, detergents, industrial wastes, DDT, etc.	Instance of cadmium, lead and mercury poisoning in human beings. Cause neurological impairment and even death.
5. Inorganic Minerals and other Chemicals like asbestos and acids, etc.	Asbestos produces lung cancer. Acids cause allergies, ulcers, skin diseases. Also have catastrophic impact on fish and aquatic life.
6. Radioactive Elements	
Wastes from nuclear power plants and nuclear fuel reprocessing plants are packaged and buried under water/oceans.	Potentially hazardous in the event of leakages and induce radiation-related illness.
7. Sediments	
Soil and mineral particles washed into streams or water.	Reduce amount of sunlight available for marine plants, cause clogging of filter plants.
8. Thermal Pollution	
Water used for cooling in electric, thermal, nuclear power plants, when recirculated in water bodies.	Can raise lake water temperature by 70–100°C, aggravate diminution of dissolved oxygen, kills marine fish and plant life.

methods not only to control water pollution but also for the purification of polluted water. But, most of these techniques are expensive and even those countries are also facing difficulties in controlling water pollution. Apart from technological methods, there are other measures which are useful in minimizing water pollution. Some of them are as follows:

1. All domestic and municipal effluents be drained in water bodies only after proper treatment.
2. As far as possible domestic waste water be used for irrigation, it is very useful for horticulture.
3. Similarly, sewage can be mixed with soil, it will act as a fertilizer.
4. The drinking water sources and its nearby area be kept clean and regular cleaning be done.
5. There must be strict regulations for industrial effluents, that they must not discharge highly toxic water into water bodies. Every industry should develop its own effluent treatment plant.
6. Use of pesticides in agriculture should be limited and only standard quality pesticides should be used.
7. There is a need to conserve several water purifying organisms.
8. In the Indian context, there should be a complete ban on the disposal of dead bodies in rivers. The bank of water bodies are often used as public latrines, this should be strictly checked.
9. Algae and other water-borne vegetation should be cleaned regularly.
10. Chemicals such as potassium permanganate be spread regularly in order to protect water from microorganisms.
11. Under present practices, pure, clean, drinking water is used to flush toilets and irrigate lawns. There are two possibilities to check this loss – first, waste water from bathroom and sinks be recycled and be used for irrigation and flushing of toilets and second, separate water supply system be provided for drinking water and for other uses.
12. The most important aspect would be to educate people not to pollute water. This awareness will help in solving this problem to a great extent.

(c) Noise Pollution

Noise is any sound that is not wanted. It is one of the most common forms of atmospheric pollution. Sound is caused by the vibration of molecules, while noise radiates from vibrating surfaces. More and more noise is the creation of modern civilization and has now become a major environmental pollutant, specially in urban areas. In 1972, the UN Environment Conference at Stockholm, accepted noise pollution as a problem, which needs proper control.

Noise pollution is an environmental problem that has only recently received much attention, yet it has been in the making for decades. The perception of noise as a problem is different for different people. Noise at certain levels might be an aggravation or nuisance for some, while at higher levels, noise might become a more widespread or general aggravation. At still higher levels, noise might fall within 'discomfort range' and at even higher levels, it may lead to physical and psychological damage.

The most popular measure of noise level is the 'decibel' measured by an instrument known as decibel metre. A sound between 0 and 1 decibel is about the weakest that the average human can hear. For testing purposes, 0 decibel is considered to be the threshold of hearing. A whisper is about 20 decibels and an average speaking voice is about 60 decibels. The loudest sound that a person can stand without discomfort is about 80 decibels. Automobile horns may reach 90 decibels and a jet airplane may have an intensity of about 140 decibels. In most of the countries, maximum limit fixed for noise is 75 to 85 decibels. According to scientists, noise above 90 decibels for a long time may result in deafness.

Sources of Noise Pollution

The sources of noise can be divided into two categories, viz. natural and artificial.

Natural sources are associated with natural phenomena like lightning, thunder, volcanic eruption, earthquakes, sound of the ocean waves, etc. These reasons are local or regional and generally their impact is very limited and not very harmful. On the contrary, noise produced by man is much greater and harmful.

Artificial sources are responsible for noise pollution. The sources of noise pollution in an industrial society are manyfold. The main sources in cities and other highly developed countries are noise associated with transportation, automobiles, motorcycles, rail engines, buses, fire brigade, police cars, ambulances, air planes, industries, high pitched musical instruments, etc. During religious or domestic functions and in political meetings during elections, noise pollution often increases due to the use of electronic medium of sound.

Effects of Noise Pollution

The most notable effect of noise pollution is on hearing. Violent noise can cause temporary or permanent impairment of hearing, thus a cause of deafness. Continual noise can lead to gradual decline in auditory acuity and eventual deafness. Acute damage occurs to the ear drum when exposed to very loud, sudden noises.

Noise causes several undesirable effects also. Damage to hearing and loss of sleep are only two of the most obvious results. Noises produce irritability and a feeling a fatigue and may reduce a worker's efficiency. There is evidence that noise is one of the major cause of stress and anxiety. It also has its impact on blood pressure and is a cause of other cardiovascular diseases. The rate of heart beat may be affected by noise. A recent report indicates that blood is also thickened by excessive noises. Eosinophilia, hyperglycemia, hypokalaemia and hypoglycemia are also caused by a change in blood and other body fluids due to noise. Noise also causes headaches and irritability. The effects of noise on the foetus are not fully known. Medical scientists have noted that an unborn child will move and kick when there is a loud noise. Some of the studies indicate that children born in areas of high noise have some defects in comparison to the children born in peaceful areas.

Control of Noise Pollution

Since noise pollution is created by man, it can be controlled by adopting certain measures. Some of the measures are as follows:

1. Noise producing industries should be located away from the residential areas.
2. Inside industries, proper arrangements to minimize noise be done by constructing sound proof walls and also to provide such instruments to workers which can protect their ears from noise.
3. Old machines often create more noise, therefore, all the machines should be well maintained and replaced if necessary.
4. The automobile horn should be designed in such a way that the noise it produces may not be harmful.

5. Use of horn should be minimum and pressure horn should be banned as has been done in many countries.
6. The noise created by musical instruments and other indoor equipments can be checked by individuals in their own interest.
7. The noise created by railways can be checked by construction of ballastless rail tracks.
8. Special arrangements be done to check noise near air bases. Control of aircraft noise requires several changes, which should be done.
9. Dense tree plantations around or nearby areas of noise can also reduce its intensity.
10. Every government has enacted certain laws to control noise pollution. The basic need is the proper implementation of these laws and regular supervision. In countries like India, the problem of noise pollution is still in its primary stage therefore, measures to control this problem are necessary.

(d) Soil and Land Pollution (The Problem of Solid Wastes Disposal)

Land pollution can be defined as the 'changes in physical, chemical and biological conditions of the soil through man's intervention or misuse of land that resulted into degradation in quality and productivity of the soil'. The problem of land pollution differs from water and air pollution in the respect that the pollutants remain in place for relatively longer periods. Rapid urbanization with the consequent increase in buildings has resulted in the reduction of lands for the wastes to be disposed. Dumping of industrial and municipal wastes causes leaching by toxic materials which seep into the soil and affect the ground course. Agricultural practices introduce pesticides, fertilizers and manures to the land resulting in both biological and chemical contamination. Thus, land and soil pollution is increasing day by day with hazardous materials and microorganisms entering the food chain or water, and are being consequently ingested by man. Therefore, there is an urgent need not only to control soil pollution but also for its proper management.

Sources of Land Pollution

The soil and land pollution is the result of several sources, which can be categorized under following heads:

1. Domestic and municipal wastes,
2. Industrial and mining wastes,
3. Agricultural wastes,
4. Radioactive materials, and
5. Biological agents.

Domestic and Municipal Wastes

One of the main causes of land and soil pollution is the growing quantity of domestic and municipal wastes. Household garbage includes food scraps, old newspapers, a variety of plastic items, bottles, discarded papers, wood, lawn trimmings, glass, canes, old appliances, tyres, worn out furniture, broken toys, and a host of other items. The total quantity of solid waste is large and increasing. In United States, municipal solid waste averaged about 1.2 kg per person per day in 1920. The quantity rose to 2.3 kg in 1970 and 3.6 kg in 1980 and now its quantity is more than 4.5 kg. Urban wastes comprise of both commercial and domestic wastes including dried sludge of sewage. In general, all the urban solid wastes are referred to as 'refuse'.

The amount of solid waste generated is directly related with prosperity. In contrast to the 3.6 kg per person per day in the United States, residents of Australia produce only about 0.2 kg per day.

The dumping of domestic and municipal wastes is a serious problem in cities because of its impact on environment and public health. Solid wastes may or may not cause diseases in man but are hazardous to health. Diseases such as dysentery, diarrhoea, plague, malaria and numerous others are the result of the indiscriminate dumping of wastes.

Industrial and Mining Wastes

The disposal of industrial solid wastes is the major source of soil pollution by toxic chemicals. The industrial wastes are mainly discharged from coal and mineral mining industries, metal processing industries and engineering industries. They contain toxic metals such as lead, copper and chemicals having acids and are responsible for soil pollution. It has been reported that about 50 per cent of raw materials ultimately become waste products in industry and about 15 per cent of it are toxic. The chemicals discharged from the industries often enter the surface or groundwater or poison the soil or crops.

The expansion of mining activities in many countries of the world has now become a main cause of land pollution due to loss of soil and destruction of land. Mining has become a threat to the environment because it leads to the huge quantity of waste generally not useful to man and its reprocessing is not economical. Mining operation produce about 1.35 billion tonnes of debris each year.

Agricultural Wastes

Agricultural practices also pollute the soil. According to an estimate agricultural activites produce more than 1.8 billion tonnes of waste each year. About three-quarters of this is manure. Much of this manure is piled in dumps where it pollutes streams and waterways. Yet, at the same time, farmers across the continent are suffering from worn out and depleted soils. Other agricultural wastes include branches and slash left over from logging apart from animal wastes. In addition to fertilizers and pesticides, soil conditioners and fumigants are used in agriculture. Organic compounds containing lead, mercury and arsenic, when applied to a land will accumulate on the soil permanently and introduce these toxic metals into plant products.

Radioactive Materials

The radioactive wastes produced by nuclear testing laboratories and industries reach the soil and accumulate there. Wastes from nuclear reactors contain ruthenium-106 and rhodium-106, iodine-131, barium-140, lanthanum-140, cerium-144, etc. All the radionuclides deposited on the soil emit gamma radiations and are harmful to the soil as well as for plant growth.

Biological Agents

The excreta of humans, animals and birds is also a source of soil pollution by biological agents. Digested sewage sludge which is used as a manure also causes soil pollution. In the developing countries, intestinal parasites constitute the most serious soil pollution problems. Faulty sanitation, waste water and wrong methods of agriculture also induce soil pollution. There are three groups of pathogenic organisms that pollute the soil, viz. 1. pathogenic organisms excreted by man, 2. pathogenic organisms excreted by animals, and 3. pathogenic organisms found in contaminated soil.

Harmful Effects of Land and Soil Pollution

The following are the harmful effects of the land and soil pollution:

1. Land and soil pollution is responsible for loss of fertility and productivity of soil.
2. The decomposition of the various types of waste material causes harmful gases and bad smell, which not only pollute environment but is also harmful if mixed with chemicals.
3. The municipal and domestic waste is often discharged in water bodies, thus responsible for water pollution.
4. The dumping of waste in oceans is a common practice and a main cause of imbalance of the marine ecosystem.
5. Particles in the sewage may clog the micro holes of the soil and also destroy the microorganisms necessary for the soil.
6. Industrial and chemical wastes are also responsible for the pollution of underground water.
7. Those bacteria which are transmitted from man to soil infect man causing dysentery, cholera, tuberculosis, typhoid, paratyphoid fever, etc.

In fact, land and soil pollution is one of the main cause of air and water pollution, therefore its control and proper management is necessary not only in developed countries but also in developing countries.

Control of Land and Soil Pollution: Methods of the Disposal of Wastes

Land and soil pollution can be controlled by adopting measures for the disposal of solid wastes, which unfortunately involves large finances and technology. Prior to disposal, the wastes have to be collected from the places by trucks and taken to the disposal area. It is estimated that about 80 per cent of the total cost of solid waste management is spent for collection only. The method used for the disposal of waste are open dumps, ocean dumping and sanitary landfills.

Dumping is a popular and an inexpensive means of solid waste disposal. Open dump is the least acceptable method but prevalent in most of the developing countries, including India and is a cause of soil as well as general environmental pollution. In this method, waste is collected and hauled to a site where it is dumped on the ground. Organic matter rots or is consumed by insects, rats or hogs. In some communities, the pile is set afire in the evening to reduce the total volume and the odour. All this creates serious ecological problems. The dump itself is a potential source of disease. Rain erodes the dump, and the polluted water flows into nearby rivers and groundwater reserves.

Sanitary landfill is commonly used for final disposal of solid wastes. This is cheaper than the other disposal methods and have the advantage of avoiding acute pollution problems associated with discharging wastes into waterways or polluting air from incineration. But, the main disadvantage over open dumps is in the aspect of public health. Leaching by surface water or groundwater may contaminate streams and other water supplies. The availability of land for the dumping of the growing volume of wastes is also a problem.

Ocean dumping of wastes is practised by many coastal cities. It was considered to be the favourite disposal method under the prevalent notion that the ocean is an inexhaustible sink. We are now learning that the ocean does not have an infinite capacity for covering, holding, absorbing or decomposing materials, nor the capacity to handle anywhere near the load of pollution that would result if the rate of ocean disposal continues, as in the past. There are seven major categories of waste – harbour dredging, industrial wastes, municipal sewage sludge, refuse and garbage, construction and demolition debris, military explosives and chemicals, and miscellaneous wastes like agricultural land run-off, pesticides, fertilizers, sewage and garbage from vessels, oil spills, etc. Now, the environmentalists have warned that ocean dumping is neither a safe nor a desirable method of the disposal of waste. In 1970, the Council on Environmental Quality recommended a series of drastic actions to control ocean dumping because it upsets the ecological balance of regions of the sea.

Incineration is a method through which the volume of waste can be reduced by 60 per cent and it also reduces the public health problems. This method is adaptable over a wide range of capacities from small domestic incinerators to large centralized municipal plants. It can handle a mixture of garbage and rubbish and the physical nature of the residue – called clinker – aids in the disposal of the waste, although it aggravates the problem of air pollution.

Chemical processing of solid wastes is also applied but it is costly and technically not feasible for poor countries. But due to prospect of energy production and recovery of certain usable materials, the salvage of some wastes by chemical methods is now considered economically feasible. The reclamation of non-ferrous scrap metals has long been an established industry.

Composting municipal refuse to convert it into a fertilizer and soil conditioner is appealing not only because it appears to be a good way to recycle the resources in solid wastes, but also because of the beneficial nature of the compost. This may be considered as a useful method for the disposal of municipal refuse, specially in developing countries.

New methods of waste disposal that are under trial in various parts of the world are:

- The transformation of the organic content into sugar or proteins.
- The heating of organic refuse under anaerobic conditions to convert it into useful gases such as methane, which is usable as fuel, or into a liquid product.
- The compression of refuse into building blocks that can be sheathed with more durable materials. One such plant is in operation in Japan.
- The compression of refuse into briquettes that can be used as fill. A mixture of fly ash, dried sewage sludge, incinerator residue, and river and lake dredging is under study.
- The use of degradable plastic bags instead of trash cans in the collection and transport of municipal refuse.
- The transport of refuse as a liquid slurry in pipelines, a method now under study.
- Recycling of waste material.

Apart from refuse and recycling of waste, there are methods through which waste can be utilized properly.

Energy from Refuse

In many metropolitan areas of developed countries, the garbage is burnt in a well-engineered furnace. The heat from the fire is used to boil water and produce steam. Then, the steam is sold for industrial use. But there are problems in this type of burning also, such as, wet garbage and food scraps are difficult to burn and burying PVC products produces hydrogen chloride gas, which reacts with water to produce a strongly corrosive liquid, i.e., hydrochloric acid. In spite of these difficulties, incineration may become profitable in the future because:

1. Increasingly large quantities of dry paper and cardboard have appeared in refuse, thereby increasing the fuel content of the trash.
2. The price of fuel has skyrocketed and therefore, the value of steam has also gone up, thus it will be more economical day by day.
3. The rising cost of land has made it harder to find adequate sites for landfills, thus demanding alternative solutions and burying of refuse is the most suitable alternative.

New Approach

The above mentioned methods have proved quite useful for waste management. But, nowadays, there is more emphasis on people's participation in waste management. There is a growing adoption of the polluter-pays principle, i.e., NIMBY and NIABY, which is as follows:

The polluter-pays principle, NIMBY and NIABY: The public is becoming better informed on environmental issues, and may object to real or imagined pollution and waste hazard or nuisance. Special interest groups or the public can adopt a 'not-in-my-backyard' (NIMBY) attitude. NIMBY is often invoked when authorities are proposing to site a hazardous or 'nuisance' waste or pollution treatment plant. Some proposals are opposed by environmental activists and NGOs whenever they are to be sited (for example, deep underground storage of high-level nuclear waste) – a 'not-in-anybody's-backyard' (NIABY) response.

Apart from the polluter-pays principle, over last few decades, pollution and waste management have been aided by:

- A trend towards more proactive planning and management;
- Increased release of information on pollution and waste risks, as a result of government, NGO, media, and international agency activity;
- Improved legislation;
- Development of better policing;
- Spread of better environmental quality standards more widely applied;
- Better methods and equipment for monitoring and assessment, etc.

(e) Radioactive Pollution

Nature and Sources

Radioactive substances are among the most toxic materials known. The magnitude of their injurious effects is in a range that might be said figuratively to represent a 'quantum jump'. H.P. Jammet, Chief of Atomic Hygiene and Radio Pathology, French Atomic Energy Commission, France, defines radioactive pollution of the environment as 'the increase in natural background radiation, emerging from the activities of man involving the use of naturally occurring or artificially produced radioactive materials. The atmosphere screens out much of the sun's radiation, including most of that which would be lethal to life'.

Naturally occurring, radioactive elements are present in the rocks, water, air, and in all living organisms. The composite of all forms of natural radiation that we are exposed to, is called background radiation. For millions of years, background radiation remained at a relatively constant level. Recently, however, the radiation that we are normally exposed to has increased and can be expected to increase further. Man-made radiation includes that from X-ray machines, radioactive fallout, nuclear reactors, research laboratories, industrial and medical uses of radioactive materials, and radioactive wastes from the production of radioactive materials.

Environmental radiation may be divided into two types: 1. natural, and 2. man-made radiations.

Naturally occurring radiations are: (i) cosmic radiations from the outer space reaching the earth's surface, and (ii) terrestrial radiation from natural radio-isotopes present in the earth's crust. Cosmic ray bombardment of the atmosphere continuously produces a few radioactive materials of short half-life. The half-life of a radionuclide refers to its period of radioactivity and has a bearing on the length of time that a radioactive substance might remain active as an environmental pollutant.

Man-made radiations originate from the activities of man involving the use of radioactive materials. They are used for the production of nuclear weapons, nuclear fuel and for the production of electric power. One of the most harmful uses of radioactive material is in the nuclear weapons, specially in atom bombs. The world has already seen the nature of destruction as well as impact of radioactivity on man and environment during the Second World War when atom bombs were dropped over Hiroshima and Nagasaki, the two Japanese towns. During the last forty years, there has been a number of nuclear weapon testings either in the sea or underground, resulting in the increase in the level of radiation of the environment. A study revealed that the radioactive materials which cause such explosions are fission products such as strontium-90, caesium-137, iodine-131 and unused explosives. These elements either settle down in the soil or water or remain suspended in the atmosphere, thus becoming a major cause of radioactive pollution.

Nuclear fuels used in operation of reactors also contribute to pollution. The heat liberated during fission and activation process is converted into electricity. During this process, two types of wastes are formed: 1. the fission products remaining in both the primary and secondary fuels, and 2. extraneous activation products in the coolant. The disposal of radioactive or nuclear wastes is a great problem and a cause of environmental pollution. It includes remaining metal products at the site of mines, fission products and activation products. These are of three categories, viz., low level, intermediate level and high level. Among these, high level products remain in the environment for several hundred years. Similarly, leakage in nuclear reactors often become the cause not only of environmental pollution but also of the death of hundreds of people as happened in the case of leakage at Chernobyl Atomic Reactor in Erstwhile USSR on 26th April 1986.

Effects of Radioactive Pollution

Different forms of radiation have different effects. Visible light and infrared heat rays and certain kinds of radiation are generally beneficial. Some of the other forms of radiation, for example, X-rays are more energetic and are therefore, biologically injurious because of their destructive action on cells and tissues. Because these highly energetic forms of radiation tend to split substances, including living matter, into ions, they are called ionizing radiation. The other forms include electromagnetic radiation and particulate radiation. Ultraviolet radiation is highly injurious to the tissues of some organisms. It is well recognized that sun-burn is caused largely by the ultraviolet content of the sun's radiation and that excessive exposure can cause serious injury to both the superficial and deep layers of the skin.

The environmental hazards from fission reactors include: 1. the possibility of a serious accident that releases a large amount of radioactivity, 2. the hazard of even the usual small radioactive emissions with normal operations, and 3. the pollution caused by radioactive wastes. The fate of a radioactive contaminant is also dependent on ecological conditions. Radioactive pollution causes deposition of the contaminants on ground and surface waters. The water consumed by plants, thus acts as a medium for radioactivity. This contamination is passed on to vegetation and animals through the nourishment from soil and water.

Man is considered to be the ultimate sufferer through reactions and interactions. Radioactive particles present in air have a direct impact on man's respiratory system, while indirect contamination occurs by consumption through the food chain. The ill effects of radioactive pollution are numerous. Radiation can cause cancer, abnormal births, skin diseases, and mutations in man. Chronic exposure to radiations leads to leukemia in an individual and even affects an unborn child in the womb. The radiations from medical X-rays, colour TV, luminous dials of clocks and watches, X-ray fluoroscopes are also contributing to the risks. As described earlier, any leakage of radioactivity from reactors is not only a health hazard but a cause of death and other disorders to thousands of people. Similarly, nuclear testing in sea waters, not only creates disorder in the oceanic ecosystem but becomes a cause of the destruction of marine life.

Control of Radioactive Pollution

In spite of certain harmful effects of radioactivity, nuclear power and other uses of radioactive elements have become an integral part of modern development. Many of the environmentally conscious countries in the world depend on nuclear power significantly for their electricity generation. At present, 17 per cent of the world's electricity is generated through nuclear sources. Nuclear source is considered clean, compact and concentrated. It is clean because its impact on the environment is minimal compared to the other energy generating sources. To make it environmentally safer still, there is a need for monitoring and prevention measures.

Monitoring of radiation is very necessary for the control of radiation. For this, sampling as well as its proper analysis is necessary. The samples of air, water and soil have to be collected regularly and their analysis both manual and instrumental must be done, not only to know the actual level of radioactive pollution but also for its control.

Prevention and control measures are to be adopted to control radioactive pollutants. All measures should so aim that the level of radioactive pollution should not exceed the maximum permissible limit. The industrial wastes having radioactive elements should only be discharged after proper treatment. The high activity wastes should not be discharged. Now, a few techniques have been developed through which these wastes can be used if treated properly. The disposal of the wastes into the sea is also not a safe method but if nothing is possible, these wastes should be sealed in concrete-filled steel drums and discharged into depths of 1,000 fathoms or more.

The emission of radioactive pollutants should also be controlled. In uranium mines, wet drilling should be employed. Closed-cycle systems with gaseous coolants of very high purity may be used in the case of nuclear reactors to prevent extraneous activation products. There is a need to control radiation hazard for workers and officials working in reactors or other places having danger of radioactivity. There is a need for regular monitoring and clean-up of surface, careful washing is necessary; apart from this, proper kit to protect from radiation effects and health care of workers are also necessary. What is more, personnel at power stations must be trained in radiation protection and radiation hazard control measures. The UN and other agencies of the world are in favour of total ban on nuclear weapons but very little progress has been made in this field. Let us hope that the time will soon come when our world will be free from the danger of atom bombs and other nuclear weapons.

Energy Crisis and Conservation of Energy Resources

Almost 90 per cent of the world's current energy supply is based on fossil or mineral sources, oil, coal, gas and uranium. Only 10 per cent is from renewable sources, especially hydropower and biomass, while the share of solar and wind energy is below 1 per cent.

It is undisputed that the structure of power supply involves three problems: if consumption rates do not change, reserves of non-renewable energy will largely run out in next one or two generation; the waste products caused by consumption in the water, the earth and in the atmosphere threaten mankind and the environment; and finally the cost of this shortage and the effects increase day by day and according to an estimate the oil and gas will be in shortage in mid 21st century and coal reserves will also fall short after 2200 AD. Even today, there are problems in coal mining. The world supply of natural gas is also limited.

Nuclear power, which amounted to over 5 per cent of global energy production accounts for very limited use and after the Chernobyl accident more technological precautions are needed. At present, there is a need for adopting methods for conservation of energy resources and also to develop and use more and more renewable energy sources.

The conservation methods should be applied for coal, petroleum and natural gas in order to preserve these resources, so that their use may be possible for a longer period and their environmental effects can be minimized. The coal mines have become deeper in most of the countries, therefore, new techniques should be applied. The new method like gasification may help in deeper mining operations. The coal should be classified and cleared at production centres itself. The use of secondary coal products like coaltar, ammonium sulphate, gas, light oil, benzol, sulphur, etc., should be made. A better transportation and marketing system as well as research in use and mining of coal should be done.

The conservation of petroleum or mineral oil is the need of the time because the availability of petro-reserves is limited and the use of oil is increasing day by day, especially in the field of transportation. The following steps may help in conservation of petroleum:

(i) Reduction of the consumption of petroleum,
(ii) Development of such engines which consume lesser petroleum,
(iii) Intensive survey for new resources,
(iv) Exploitation of oil resources from ocean floors,
(v) Correct estimation of available resources,
(vi) New techniques for oil drilling and refining should be developed,
(vii) To reduce petroleum consumption in certain areas such as, for military purposes,

(viii) to develop alternative source of energy of automobiles, etc., petro-products, and

(ix) research institutes working in this field should intensify their work.

The energy crisis can be solved to some extent by proper use of renewable sources of energy. The renewable energy sources include solar energy, wind energy, tidal energy, hydropower, ocean energy, and geothermal energy. Renewable energies are also called the energies of the future, because in principle they are also inexhaustible and neutral in terms of CO_2. The renewable energy sources utilized by man come from three primary sources:

- isotopic disintegration in the earth's interior,
- thermonuclear conversion (fusion) in the sun, and
- the movement of planets, combined with mass attraction.

The greatest of the renewable energy sources is solar energy or solar radiation reaching the earth's atmosphere year after year. Every 29th second, the solar energy that falls on our planet is equivalent to human energy needs for one day at the 1980 consumption level converted to heat through absorption. This can occur through passive systems, i.e., through constructional measures for collecting, storing and distributing solar energy which renounce thermal installation to a considerable degree or through active systems, i.e., collectors or absorbers. The advantages of using solar energy are: limitless supply, no air pollution, no water pollution, no harmful wastes, no possibility of a large scale explosion or disaster. There is a need to evolve cheaper technology to harvest solar energy. At least for tropical regions, it is the greatest source of energy.

There is no need to stress the importance of *water energy* or hydropower. Water energy is already in use in developed countries, but in developing countries vast energy potentials exist.

There is enough *wind energy* available to supply the world's energy needs, and it is technically feasible to build windmills capable of producing electricity. These can be build either as small home-sized units or as large central generators. Besides the conversion of rotation energy into electrical energy, the conversion of rotation energy into mechanical energy also finds application in pumping of water, etc.

Ocean energy refers to a number of different renewable energy sources – tidal energy, wave energy, ocean currents and ocean heat.

Today, the significant attempts at harnessing the energy of the ocean have been related to tidal power. In many costal regions, the flow of the tide naturally funnels through narrow entrances into bays and estuaries, and strong currents can be used to generate power.

Energy derived from the heat of the earth's crust is called *geothermal energy.* In various places on the globe, hot springs, geysers and underground steams are available, which can be used for power generation.

Bioenergy can be produced by organic compounds through chemical and biological processes. The combustion, gasification and liquification helps in production of gas which can be used as a source of energy. The biogas plant has now become a common feature in rural India.

Viewed in the long-term perspective utilization of renewable energy sources is of major significance, in terms of both the conservation of the reserves of fossil fuels and the reduction of pollutant emission caused by the conversion of energy from coal and oil gas. The renewable energy sources have good chances for the future energy supply. How large their contribution will be, depends upon the question of how safe, economical and environmentally friendly each particular 'new' energy is.

Adverse Effects of Insecticides and Pesticides

The excessive use of chemical pesticides and insecticides has resulted in degradation of environment which has become a threat to human health. Through out the ages, farmers have had difficulty in protecting their crops from small animals and disease organisms. At times, pest species have boomed and brought diseases and famine. But most of the time, natural balance has been maintained, and humans have lived together with insects in reasonable harmony. Nowadays, a great debate over the use of insecticides in agriculture is going on; on the one hand, some scientists claim that more production can only be possible through control over pests and weeds, while on the other hand, some disagree and feel that chemical pesticides are harmful to the environment and actually increase pest problems.

The modern pesticide era started in about 1940 when a chemical called DDT was discovered to be a potent insecticide. DDT was cheaper and more effective than the previously known chemicals and led to dramatic early successes. But, this miracle chemical had fallen from grace and on 1st January 1973, all inter-state sale and transport of DDT was banned in USA; ban was also imposed on several other pesticides like aldrin, clieldrin, etc. Most of the insecticides are broad spectrum poisons and their use has led to disruption of ecosystems. Because these materials are not biodegradable, they persist for long periods in the environment. In some soils, DDT and aldrin have been detected 15 years after a single spray application. Such persistence leads to a series of several ecological problems. All insecticides have had an adverse impact on human health and since the mid-1940s, many thousands of people have been sickened or have died from severe pesticide poisoning every year.

Although the use of DDT in US and other developed countries was banned in 1973, it is still being manufactured and exported for use in other countries, specially to Third World countries. Thus, long-term global contamination continues. What is needed is the use of organic fertilizers and natural system of control through certain bacteria. Pests can be controlled without killing them directly. There is a need for integrated pest management and evolving scientific methods to control pests and weeds without adverse effects on the ecosystem. In addition, government restrictions may serve as a drive to develop other, more environment friendly system of pest control.

Pollution of the Oceans

For many years, people were relatively unconcerned about the pollution of the oceans because the great mass of the sea can dilute a huge volume of foreign matter to the point where it has little effect. In recent years, this attitude has changed because pollutants are often added in such relatively high concentrations that environmental disruption does occur before the pollutants have a chance to disperse. In addition, the quantity of pollutants dumped into the ocean has grown so large that some scientists fear that its global effect may be significant.

Oil spilling in oceans is a major problem. Between 1996 and 1983, there were more than 500 tanker accidents that involved oil spills, and altogether, more than one million tonne of oil was released. During the Gulf War, a huge amount of oil spilled in nearby seas caused great damage to the ocean life. Apart from accidents, seepage of oil is also common. Sometimes thousand of tonnes of oil spills out of the tankers by large leakage as happened in October, 1997 near Singapore. Even tanker captains often clean the oil holds of their ships in ocean waters.

Offshore drilling operations also contribute their share of oil to the sea. The largest spill so far started on 3rd June 1979, at an offshore oil well owned by Pemex, the Mexican National Oil Company.

The total quantity of oil that finds its way into the sea each year is very large. There is a tendency to dispose of high poisonous chemical wastes including chemical warfare agents, pesticide residues and radioactive substance into the oceans. The sealed drums are usually dumped into the sea. All these dumping activities have had an adverse effect on the entire ecology of the oceans. During the 1972, Stockholm Conference on Human Environment, a study entitled, '*The Health of Oceans*' was released, which has attracted the attention of world leaders but till now nothing substantial has been done in this field. The blasting of atom bombs under ocean waters is the greatest threat to the ocean environment, which must be banned immediately by all nations.

Floods, Droughts and Famines

Floods and droughts both are natural hazards and are related with rainfall. If there is heavy rain in the catchment area of a river, floods will occur. On the other hand, if there is less rain due to failure of monsoon, droughts will occur. In both the situations, their impact on environment creates problems and also affects the regional ecosystem.

Flood is a hydrological phenomenon that occurs in any drainage basin. Its magnitude and frequency varies primarily on the precipitation conditions in the catchment area. Most of the river valleys are subjected to varying degrees of flood incidences, either creating minor damages or devastating havoc at times. Such calamities of floods pose a serious problem to the whole environmental set-up of a region, bringing about some phenomenal changes in the physical environment, ushering in accelerated erosional, transportational and depositional activities of the river with the consequent effects on the soil. The devastating effects of floods can be seen on vegetation, agriculture, industry, and settlements, sometimes causing loss of human and animal lives and of materials also.

The main cause of flood is excessive rainfall but its occurrence is also due to earthquakes, deforestation in the upper catchment areas, failure of dams, excessive snow melting, etc. The occasional heavy rainfall due to cloudburst is also a reason for floods, as we have seen on 16th and 17th July 2013 in Uttarakhand which caused heavy destruction not only to men and material but also to the environment.

In India, the problem of floods and challenge to the environmental setting is of gigantic magnitude, because of: (i) typical seasonal rainfall, and (ii) erratic nature of tropical cyclones and rainstorms. The following are the areas which may be termed as flood-prone areas:

- Upper catchment and piedmont zone of the Himalayan river basins,
- Ganga river valley – upper, middle and lower deltaic zone,
- Brahmaputra valley and its tributaries,
- Deltaic region of Mahanadi, Godavari, Cauvery and Krishna rivers, and
- Narmada and Tapti basins.

The common measures of the flood control are as follows:

(i) Control of erosion in upper catchment areas,
(ii) Construction of dams and reservoirs at suitable sites,
(iii) Prevention of overflow in channels,
(iv) Construction of multiple channels for distribution of the excess river water, and
(v) Development of shelters in flood-prone areas.

The flood problem should be taken up with more intensive preventive measures, in order to avoid environmental problems.

Drought is also a natural hazard which has its impact on the regional ecosystem. The area where droughts become severe, they cause famines, which are devastating for man, animals and also the environment. Generally, droughts occur when rainfall is less than 25 to 50 per cent of the average. If it is below 50 per cent of the average, famines occur. In 1899, India faced the severest famine known as *Chappan-ka-Akal*. A similar famine also occurred in 1917. In both these famines, several lakhs of people lost their lives. In 1992–93, Ethiopia also suffered from the grips of a bad famine, in which more than a lakh people died. Famine conditions are common in Rajasthan, Gujarat, Haryana, Madhya Pradesh, Odisha, and Andhra Pradesh in India.

Environmental Degradation by Volcanoes and Earthquakes

The eruption of volcanoes is one of the most dreadful natural phenomena. The molten rocks, boiling lava, hot gases penetrate out of that part of the surface which is comparatively weak. When this material comes out of these spots, an explosion takes place with a thunderous noise; and molten rocks, lava and gaseous clouds rise up towards the sky. All this results in a threat to the environment as well as to life.

Volcanism is not common throughout the world but is limited to the circum-Pacific belt extending from Andes in South America, through the Cascades to the New Zealand. There is also an important belt in the Mediterranean, which includes the active volcanoes of Italy.

A volcanic explosion poses a great threat to environment. On an average, a volcano, in an explosion, ejects about 150 cubic km of material. This material is of three types: solid material in the form of small and large pieces of rocks, ash particles and also pieces of volcanic bombs which shoot into the sky; the liquid material called magma comes out of the volcano, also known as lava; and a huge amount of gaseous material also comes out of a volcano, i.e., steam, carbon dioxide and sulphur which in vapour form produce poisonous gases when they combine with other gases. Gases like chlorine, fluorine, etc., are also emitted in large quantities.

The eruptions of volcanoes and lava flows are environmental hazards of the severest type, often taking a heavy toll on plants and animal life. Loss occurs from sweeping clouds of incandescent gases and suspended matter that descend over the volcano slopes like great avalanches. When Mont Pele burst in 1902 in the West Indies, the burning material ejected out of it moved in the form of a cloud over the city called St. Piere and only two people out of a total of 30,000 population survived. The explosions of Etna and Vesuvious in Italy destroyed many beautiful cities. In 1943, when Paracutin, 320 km West of Mexico City exploded, it created a 167 metre high mountain. In 1991, volcanic explosion in Japan and Philippines damaged a lot of property, caused loss of human life and also polluted the environment.

Similarly, volcanic activity in the Andaman & Nicobar Islands was responsible for the environmental degradation in this area.

Earthquakes are tremors which are produced by the passage of vibratory waves through the rocks of the earth. The devastation caused by earthquakes to the life of man and property, as well as to the ecosystem is very high. The earthquake belt of the world is also associated with the volcanic belt, but several other regions too have occurrence of earthquakes. Wherever earthquakes occur, apart from loss of life and property, landslides change the course of rivers, loss of vegetation and animal life, destruction of dams and development of faults and fractures in earth's surface are common and all these events lead to environmental degradation. The list of devastating earthquakes in the world is too long to be enumerated here. In India, the occurrence of earthquakes in areas of Jammu and Kashmir, Himachal Pradesh, Uttar Pradesh, Bihar, West Bengal, Punjab, Delhi, Haryana, Gujarat, North-eastern states as well as in Maharashtra, are common.

Thunderstorms, Tornadoes and Tropical Cyclones

Due to abrupt changes in atmospheric conditions, thunderstorms, lightning, hailstorm, tornadoes, and cyclonic storms occur and are the cause of environmental hazard. A thunderstorm is an intense local storm caused by strong convectional activity. Thunder and lighting normally accompany the storm, and rainfall is heavy, sometimes due to the cloudburst also. Violent surface wind may occur at the onset of the storm. The speed of the rising air in a thunder storm is about 112 km per hour. Lightning is also one of the environmental hazards that results annually in the death of many persons and livestock and also causes forest fires and building fires. One of the most destructive products of the thunderstorm is the hailstorm. Hail consists of rounded lumps of ice ranging from 0.5 to 5 cm in diameter. Large hailstones are an environmental hazard as they can be extremely destructive to crops and flimsy buildings.

A tornado is a small, intense cyclone in which the air spirals at a tremendous speed. Associated with frontal air mass interactions in mid-latitudes, tornadoes are the most violent storms. These storms are the most frequent and violent in the United States, although occurring in Australia and in other places of mid-latitudes as well. A tornado moves across the country with a wind speed of about 400 km per hour in a funnel like shape. The end of the funnel cloud may alternately sweep the ground, causing complete destruction of anything in its path.

Tropical cyclones are the most powerful and destructive types of cyclonic storm, which are also known as hurricane or typhoon. These storms develop over oceans in latitudes 8° to 15° N and S, but not close to the equator. The tropical cyclone is an almost circular storm having low pressure in which winds spiral at high speed accompanied by heavy rainfall. The environmental impact of tropical cyclones lies in the form of destruction of life, infrastructures as well as of forest and animals. In islands and coastal regions, large scale destruction of cities and their inhabitants has been reported on several occasions. Bangladesh is a country which has suffered a lot by heavy destruction caused by tropical cyclones. The death toll by tropical cyclones in Bangladesh is almost unbelievable. In 1822, the toll was 40,000 while in 1876 and 1897, it was 1 lakh and 1.75 lakh respectively. In 1970, 3 lakh lives were lost and in 1991, the death toll was 5 lakh. These storms are also responsible for change in atmospheric and environmental conditions.

Box 1.9
Environment Protection Act of Pollutions Act/ Rules in India, 1981

1. Air (prevention and control of pollution) Act, 1981.
2. Water (prevention and control of pollution) Act,1974.
3. Environment (protection) Act, 1986.
4. Noise Pollution (regulation and control) Rules, 2000.
5. Wildlife (protection) Act, 1972.
6. Hazardous Waste (management and handling) Rules, 1989.
7. Manufacture, Storage and Import of Hazardous Chemical Rules, 1989.
8. Biomedical Wastes (management and handling) Rules, 1989.
9. Municipal Solid Wastes (management and handling) Rules, 2000.
10. National Environment Tribunal Act, 1995.
11. National Environment Appellate Authority Act, 1997.

MULTIPLE-CHOICE QUESTIONS

1. What is the name of the set of many variables which surrounds all living organisms?

   (a) Hydrosphere (b) Environment
   (c) Atmosphere (d) Lithosphere

2. Which one of the following is not a component of environment?

   (a) Stratosphere (b) Atmosphere
   (c) Hydrosphere (d) Lithosphere

3. In which term all the three components of environment, i.e., atmosphere, hydrosphere and lithosphere are included?

   (a) Ecology (b) Autecology
   (c) Biosphere (d) Habitat

4. Which of the following statements is/are correct?

   1. The human working efficiency depends on climatic conditions.
   2. The physique of an individual depends on physical environment.

   (a) only 1 (b) only 2
   (c) both 1 and 2 (d) none of these

5. The term 'ecosystem' was first used by:

(a) Mobius (b) Forbes
(c) Theinemann (d) Tansley

6. Which of the following is not a terrestrial ecosystem?

(a) Forest (b) Grassland
(c) Swamp (d) Desert

7. The ecosystem comprises:

(a) all the animal species living in an area
(b) all the individual of a particular species living in an area
(c) all the organism living in an area
(d) all the living organism and the non-living components in are area

8. A biome is:

(a) A complex of communities characterized by distinctive climates and vegetation
(b) A group of plants growing in an area
(c) A region of distinct soil
(d) An area of a game sanctuary

9. The biosphere refers to:

(a) The area of land inhabited by living organisms
(b) The position of the earth, including oceans, land and the atmosphere inhabited by living organisms
(c) The atmosphere surrounding the living organisms
(d) That part of ocean having plants and animals

10. Man and Biosphere programme is sponsored by:

(a) UNESCO (b) WTO
(c) IUCN (d) World Bank

11. Which major biome has a dominance of thorny bushes?

(a) Forest (b) Desert
(c) Savanna (d) Tundra

12. A large unit of vegetation having uniformity of not only life form but also of adaptation to the environment is called:

(a) ecosystem (b) a biochore
(c) a biome (d) a forest

13. The organisms which can produce their own food are known as:

(a) Heterotrophs (b) Autotrophs
(c) Consumers (d) Decomposers

14. Which of the following is not an abiotic component in an ecosystem:

(a) rivers (d) solar energy
(c) plants (d) air

15. Who has given 'box' model of energy flow in a food chain?

(a) E.P. Odem (b) A.G. Tansley
(c) Forbes (d) P. Haggett

16. What is a complex of vegetation called when it undergoes no further change unless the climate changes perceptibility?

(a) End of succession
(b) Climatic climax community
(c) Edaphic climax community
(d) An ecosystem

17. In a chain of communities, succession of each community is a:

(a) link (b) step
(c) sere (d) climax

18. A food web is a:

    (a) A series of food chains having frequent interlinks
    (b) A circular food chain
    (c) A moderate sized food chain having many trophic levels
    (d) A long straight linked food chain

19. Examine the following statements and select the correct answer using the code given below:

    1. A large, stable terrestrial ecosystem characterized by specific plants and animals is known as biome.
    2. A boundary transition zone between adjoining ecosystems is known as an ecotone.
    3. Small plants without woody stems above the ground are known as herbs.
    4. Smaller woody plants, branching stems at the ground are known as shrubs.

    Code:

    (a) 1, 2 and 3 are correct
    (b) 1, 2 and 4 are correct
    (c) 1, 3 and 4 are correct
    (d) 1, 2, 3 and 4 are correct

20. The tallest trees in the world are found in the:

    (a) Equatorial region (b) Monsoon region
    (c) Mediterranean region (d) Temperate rain forest

21. Match list I with List II and select the correct answer using the code given below

List I (Biome & Ecosystem Symbol)	List II (Vegetation Characteristics)
(A) Needle leaf conifers	(1) Maple/Redwood
(B) Mid-latitudes mixed forest	(2) Teak/Sal

(C) Tropical Monsoon — (3) Spruce, Fir

(D) Tropical Savanna — (4) Trees with fattened crowns

Code:

	A	B	C	D
(a)	3	1	2	4
(b)	3	1	4	2
(c)	4	3	1	2
(d)	4	3	2	1

22. Which one of the following is not a characteristic of mediterranean vegetation:

(a) Hard leaf (b) Leathery leaves

(c) Even low branches (d) Well-developed roots

23. Which one of the following is the most important element of ecosystem?

(a) Ecological succession (b) Energy flow

(b) Food chain (d) Food web

24. The soil conservation implies:

(a) Using the soil in the most intensive manner

(b) Using the soil judiciously to maintain their fertility

(c) Using too much chemical fertilizers for soil fertility

(d) Using soil only for grazing

25. Excessive irrigation by canals may lead to:

(a) soil alkalinity (b) soil acidity

(c) soil erosion (d) soil conservation

26. Humus is the:

(a) Living organic matter
(b) Leaves and twigs of plants
(c) Partly decomposed organic matter
(d) Fully decomposed organic matter

27. Which of the following is a correct statement?

(a) Soil of an area is determined by climate
(b) Vegetation is determined by soil
(c) Both soil and vegetation is determined by climate
(d) Climate, soil and vegetation all are interrelated

28. Savanna is a:

(a) Tropical mixed vegetation dominated by grass
(b) Forest of tropical regions
(c) Grasslands of the deserts
(d) Temperate type of grassland having only grass

29. Cultivation on steep slopes lead to:

1. increased run-off
2. increased percolation
3. increased erosion
4. increased bacterial activity

(a) 1, 3 and 4
(b) 1, 2 and 3
(c) 1, 2 and 4
(d) 1 and 3

30. Xerophytes are the plants:

(a) grown in humid regions
(b) grown in arid regions
(c) grown in temperate regions
(d) grown in saline area

31. **Assertion (A):** Crop rotation helps in pest control. **Reason (R):** Crop rotation helps maintaining soil fertility

Code:

(a) Both **A** and **R** are true and R explains **A**

(b) Both **A** and **R** are true but R does not explain **A**

(c) **A** is true but **R** is false

(d) **A** is false but **R** is true

32. Ecological Pyramids represent the:

(a) Trophic structure (b) Energy flow

(c) Ecological succession (d) All the above

33. The ecological niche explain:

(a) environmental conditions

(b) macrohabitats

(c) microhabitats

(d) spatial nature of organism

34. Which of the following is not an ecological niche

(a) Habitat niche (b) Energy niche

(c) Trophic niche (d) Multidimensional niche

35. Which of the following is not an aquatic biome?

(a) Lentic (b) Swamps

(c) Estuaries (d) Chaparral

36. Which among the following is a characteristic of evergreen forest?

(a) The trees never shed their leaves

(b) All the trees do not shed their leaves simultaneously

(c) When the trees are leafless, grass remain green

(d) The name has noting to do with greenery as such

37. The plants growing in water are called:

(a) hydrophytes
(b) xerophytes
(c) hygrophytes
(d) mesophytes

38. Who form the first trophic level in any ecosystem?

(a) Herbivores
(b) Plants
(c) Carnivores
(d) Omnivores

39. The functioning of an ecosystem involves:

(a) Flow of energy and circulation of nutrients
(b) Increasing number of trees and bushes
(c) Increasing number of animals
(d) An increasing simplicity of relations.

40. Which of the following is not a biogeochemical cycle?

(a) Carbon cycle
(b) Nitrogen cycle
(c) Cycle of erosion
(d) Sulphur cycle

41. The richest forest resources of the world are:

(a) Tropical forest
(b) Sub-tropical forest
(c) Temperate forest
(d) Coniferous forest

42. The main problem of tropical forest is:

(a) Forest fire
(b) Deforestation
(c) Soil erosion
(d) Destruction of forest due to population growth

43. Which of the following is not an impact of deforestation on environment?

(a) increase in temperature
(b) increased rate of soil erosion

(c) increase in atmospheric humidity
(d) extinction of species

44. Which of the following is a suitable method of the conservation of forests?

(a) Forest survey
(b) National plan for forest conservation
(c) A proper ratio between forest cutting and plantation of trees should be maintained
(d) Restriction should be imposed on forest land conversation.

45. How much percentage of the land area is at the risk from desertification in the world?

(a) 20% (b) 35%
(c) 40 % (d) 42%

46. Which state of India is highly prone to desertification?

(a) Punjab (b) Rajasthan
(c) Haryana (d) Gujarat

47. Which of the following is not a proper step to control desertification?

(a) Growth of livestock
(b) Plantation of ecologically suitable plants
(c) Management of land use and farming practices
(d) Stabilization of sand dunes

48. Which of the statements given below is/are correct?

1. Forestry is 2nd largest land use in India after agriculture, covering approximately 23.57 per cent of total geographical area

2. Forestry accounts maximum in gross domestic agricultural product.

(a) only 1 (b) only 2

(c) both 1 and 2 (d) none of these

49. What could be the main reason/reasons for the formation of African and Eurasian desert belt?

1. It is a located in the subtropical high pressure cells
2. It is under the influence of warm ocean currents

Which of the statements given above is/are correct?

(a) only 1 (b) only 2

(c) Both 1 and 2 (d) Neither 1 nor 2

50. Among the following Indian States, which one has minimum total forest cover?

(a) Sikkim (b) Goa

(c) Haryana (d) Kerala

51. Which of the following is not an agent of soil erosion?

(a) Water (b) Vegetation

(c) Wind (d) Ocean waves

52. Which of the following is not a type of soil erosion?

(a) Sheet (b) Rill

(c) Wind (d) Gully

53. Which of the following river basins in India is having 'ravines'?

(a) Kaweri (b) Tapti

(c) Mahanadi (d) Chambal

54. Which are of the following is not a cause of soil erosion?

(a) Deforestation

(b) Overgrazing

(c) Intensive agriculture

(d) Over exploitation of vegetation

55. Which of the following Indian states are having serious problems of soil erosion?

(a) Assam, Nagaland and Mizoram

(b) Himachal Pradesh, Uttarakhand and Jammu Kashmir

(c) Kerala, Tamil Nadu and Andhra Pradesh

(d) Karnataka, Orissa and Maharashtra

56. The extreme case of soil erosion is known as:

(a) Run-off (b) Peniplain

(c) Landslide (d) Floods

57. Which of the following is a mechanical method of checking the soil erosion?

(a) Terracing (b) Strip-cropping

(c) Crop-rotation (d) Shelter belt

58. With references to soil conservation, consider the following practices:

(a) Crop rotation (b) Sand fences

(c) Terracing (d) Wind breaks

Which of the above are appropriate methods for soil conservation in India?

(a) 1, 2 and 3 (b) 1, 2 and 4

(c) 1, 3 and 4 (d) All these

59. IRENA (Intergovernmental Agency on Renewable Energy) Assembly was held at:

(a) Abudhabi (b) Glasgow
(c) Hyderabad (d) Addis Ababa

60. Soil erosion is responsible for:

(a) Loss of salts in the soil (b) Temperature rise
(c) Loss of soil fertility (d) All the above

61. Which of the following gases constitute the major component of the atmosphere?

(a) Oxygen-Hydrogen (b) Nitrogen-Oxygen
(c) Nitrogen-Argon (d) Oxygen-Carbon dioxide

62. Which of the following is not a cause of air pollution?

(a) Automobiles (b) Industries
(c) Use of solvents (d) Forests

63. The increase in the concentration of carbon dioxide, in our environment in last fifty years, since 1960 is about:

(a) 20% (b) 10%
(c) 14 % (d) 6 %

64. Which of the following is not a primary contributor to the green house effect?

(a) Carbon-dioxide (b) Carbon monoxide
(c) Argon (d) Methane gas

65. Consider the following:

1. Carbon dioxide 2. Oxide of Nitrogen
3. Oxide of Sulphur

Which of the above is/are the emission(s) from coal combustion zone at thermal power stations?

(a) only 1 (b) 2 and 3

(c) 1 and 3 (d) 1, 2 and 3

66. Coal is considered as the worst pollutant when burnt because it releases large amount of:

1. Carbon dioxide
2. Sulphur dioxide
3. Nitrogen oxides
4. Methane

Which of the above statements are true?

(a) 1 and 4 (b) 1, 2 and 3

(c) 1, 3 and 4 (d) 1, 2, 3 and 4

67. Excessive emission of carbon dioxide are responsible for ocean acidification; it means that:

1. Ocean waters become acidic
2. Only the surface waters become acidic
3. Surface waters only become less alkaline

Select correct answer from the code given below:

Code:

(a) only 1 (b) only 2

(c) 2 and 3 (d) only 3

68. Which pollutant is not emitted by petrol/diesel-driven vehicles?

(a) Carbon monoxide (b) Ozone

(c) Sulphur oxide (d) Nitrogen oxide

69. Which of the following statements about natural gas are true?

1. Natural gas is the clearest fossil fuel
2. Methane is the major components of natural gas

3. The natural gas is a product of tiny plants and animals of geological gas
4. Saudi Arabia has the highest natural gas reserves

Code:

(a) 1 and 3 (b) 3 and 4

(c) 1, 2 and 4 (d) 1, 2 and 3

70. The pollutants discharged from a thermal power station are:

(a) Hydrocarbon and nitrogen oxide

(b) Smoke and flyash

(c) Water vapour and carbon dioxide

(d) All the above

71. Which of the following industrial cluster was found most polluted in a study conducted by Central Pollution Board, in collaboration with IIT Delhi?

(a) Ankleshwar (b) Ghaziabad

(c) Ludhiana (d) Vellore

72. The country having highest natural gas reserves in the world is:

(a) Iran (b) Qatar

(c) Russia (d) Saudi Arabia

73. Consider the following:

1. Oxides of hydrogen 2. Oxides of nitrogen
3. Oxides of sulphur

Which of the above cause/causes acid rain?

(a) 1 and 2 (b) only 3

(c) 2 and 3 (d) All of these

74. Who first used the term 'Smog'?

(a) H.J. Sanders
(b) H.A. Das Voeux
(c) E. H. Brown
(d) K. Smith

75. At which place 'smog' was considered as a cause of death of more than thousand people in 1909?

(a) Glasgow
(b) Pennsylvania
(c) London
(d) Yokohama

76. Which of the following is a method to minimize the air pollution of the urban centres?

(a) Agricultural development
(b) Green Belt
(c) Better housing
(d) Lesser use of automobiles

77. Match List I with List II and select the correct answer using the code given below:

List I (Polluted Area)	List II (Type of Industry)
A. Greater Kochi	1. Cotton textile and dyeing
B. Nagda-Ratlam	2. Food processing and electroplating
C. Pali	3. Oil refineries, chemical and metallurgical
D. Parwanoo	4. Viscose rayon, caustic soda, dyes and distillery

Code:

	A	B	C	D
(a)	4	3	2	1
(b)	3	4	2	1
(c)	4	3	1	2
(d)	3	4	1	2

78. The most suitable measure for soil conservation is:

(a) Afforestation (b) Irrigation

(c) Crop rotation (d) Counter farming

79. Consider the following statements:

1. 2.7 per cent of the water present on the earth is fresh water
2. Majority of fresh water is stored in cryosphere of the earth, so it is not usable,

Which of the statements given above is/are correct?

(a) only 1 (b) only 2

(c) Both 1 and 2 (d) Neither 1 nor 2

80. Consider the following statements:

1. 37 per cent geographical area of India is flood-prone
2. Flood-prone areas in India has increased exponentially

Which of the statements given above is/are correct?

(a) Only 1 (b) only 2

(c) Both 1 and 2 (d) Neither 1 nor 2

81. Which one of the following statements is correct?

(a) All the abiotic resources are non-renewable and all the biotic resources are renewable

(b) All the abiotic resources are non-renewable but some of the biotic resources are non-renewable.

(c) Some of the abiotic resources are non-renewable but the biotic resources are renewable.

(d) Some of the abiotic resources are renewable and some of the biotic resources are non-renewable.

82. Which part of human body is highly affected by air pollution?

(a) Blood circulation system
(b) Respiratory system
(c) Mind system
(d) Digestive system

83. **Assertion (A):** When the absorption of sulphur dioxide exceeds a particular level, the cells become inactive and are killed resulting in tissue damage and drying of leaves.

Reason (R): Air pollution has caused widespread damage to trees, fruits and vegetables in general vegetation as a whole. Select the correct answer from the code given below:

(a) Both **A** and **R** are true and R is the correct explanation of **A**
(b) Both **A** is true but **R** is false
(c) **A** is true but **R** is false
(d) **A** is false but **R** is true

84. Which are the useful methods to control gaseous pollutants of air?

(a) Combustion
(b) Absorption
(c) Adsorption
(d) All the above

85. In which year Air (Prevention and control of Pollution) Act was introduced in India?

(a) 1974
(b) 1981
(c) 1986
(d) 1972

86. Which of the following is a pollution free source of power?

(a) Coal
(b) Petroleum
(c) Hydroelectricity
(d) Natural gas

87. Which of the following is the most chemically polluted city in the world?

(a) Linfen (China) (b) Dzerzhinsk (Russia)
(c) Sum gayti (Azerbaijan) (d) La Oroya (Peru)

88. Which of the following is not a toxic metal?

(a) Lead (b) Cadmium
(c) Gold (d) Copper

89. During Bhopal Gas Tragedy, which gas was released in environment?

(a) Sulphur dioxide (b) Methyl isocyanate
(c) Carbon monoxide (d) Nitrogen dioxide

90. Which of the following is not a greenhouse gas?

(a) Carbon dioxide (b) Helium
(c) Methane (d) Nitrous oxide

91. Which of the following is not a source of water pollution?

(a) Automobiles (b) Domestic effluents
(c) Sewage (d) Industrial effluents

92. Which one of the following is the main source of water pollution in urban areas?

(a) Agricultural effluents (b) Automobiles
(c) Municipal wastes (d) Radioactive wastes

93. Which of the following statement (s) is/are correct?

1. Polluted stagnant water is a breeding ground for parasites, bacteria, viruses, etc.
2. Synthetic organic compounds are the cause of water pollution as well a of neurological impairment.

3. Water used for cooling the thermal power plants, when recirculated in water bodies kills fish and plant life.

Code:

(a) only 1 (b) only 2
(c) 1 and 3 (d) 1, 2 and 3

94. The main cause of marine pollution is;

(a) Fishing (b) Oil spills
(c) Atomic experiments (d) All the above

95. Excess irrigation result into:

(a) Higher productivity (b) Soil erosion
(c) Salinity (d) Loss of soil fertility

96. There has been increasing incidence of floods in northern India in recent years because:

(a) The ice has been melting faster due to rise in temperature
(b) There has been increased deforestation in the catchment area
(c) The rate of silting of dams has gone up
(d) There has been an increase in rainfall

97. Flouride content in water is responsible for the disease:

(a) Ameboiosis (b) Typhoid
(c) Bone disorder (d) Cholera

98. Which is not a type of natural water impurity?

(a) Suspended impurities (b) Gaseous impurities
(c) Colloidal impurities (d) Dissolved impurities

99. What is the impact of degradable wastes when mixed with water?

(a) Depletion of dissolved oxygen in water
(b) Growth of algae
(c) Reduced amount of sunlight available for marine plants
(d) Rise in water temperature

100. Why is Damodar River considered as a highly polluted river?

(a) Due to soil pollution
(b) Due to discharge of domestic wastes in the river
(c) Due to lack of purification systems
(d) Several industries discharge their waster into the river

101. SO_2 pollution is indicated by:

1. Deschampsia (Grasses)
2. Sphganum (Mosses)
3. Usnea (Lichens)
4. Cucurbita (Climbers)

Select the correct answer using the codes given below:

(a) 1 and 2 (b) only 3
(c) only 1 (d) only 4

102. Consider the following statements about cyclones:

1. Tropical cyclones are more violent than temperate cyclone
2. Tropical cyclones live longer than the temperate cyclone

Select the correct statement(s) using the codes given below:

(a) only 1 (b) only 2
(c) Both 1 and 2 (d) None of these

103. **Assertion (A):** The pH of acid rain lies between 9 and 12
Reason (R): Acid rain occurs due to heavy air pollution.

Codes:

(a) Both **A** and **R** are true and **R** is the correct explanation of **A**

(b) Both **A** and **R** are true but **R** is not a correct explanation of **A**

(c) **A** is true but **R** is false

(d) **A** is false but **R** is true

104. Tsunamis are produced by:

(a) Shrinking of earth's crust

(b) Cyclone

(c) Submarine earthquakes

(d) Tides

105. The tallest trees of the world are found in

(a) Equatorial biome

(b) Tropical deciduous biome

(c) Temperate rain-forest biome

(d) Mediterranean biome

106. Among the following climatic factors, which one has the least effect upon terrestrial ecosystem?

(a) Temperature variation (b) Wind

(c) Conditions of sunlight (d) Availability of water

107. Consider the following statements:

1. Counter ploughing promotes soil erosion and formation of gullies.
2. Gully erosion is the removal of top soil and creation of many step-sided cuttings in a hill side.

Which of the statements given above is/are correct?

(a) only 1 (b) only 2
(c) Both 1 and 2 (d) Neither 1 nor 2

108. Among the following environmental pollutants, which one leads to the occurrence of acid rain?

(a) Carbon dioxide (b) Hydrogen peroxide
(c) Carbon monoxide (d) Sulphur dioxide

109. The Minamata incident, an example for environmental degradation is due to which one of the following:

(a) Air pollution (b) Nuclear disaster
(c) Water pollution (d) Thermal pollution

110. Among the following greenhouse gases, the emission of which one of the following is entirely anthropogenic?

(a) Choloro fluoro carbons (b) Methane
(c) Nitrous oxide (d) Hydrogen sulphide

111. What step has been taken in 1984 to clean river Ganga?

(a) Pollution survey of Ganga River was started
(b) Steps have been taken to stop industrial effluents in river Ganga.
(c) Central Ganga Authority was formed to implement Ganga Action plan.
(d) Water purification plants were sited along river Ganga.

112. Select the correct sequence of food chain in the Arctic waters from the following:

(a) Zooplankton-diatoms-fish-polar bear
(b) Diatoms-zooplankton-fish-seal-polar bear
(c) Fish-diatoms-seal-zooplankton-polar bear
(d) Diatoms-fish-polar bear-seal-zooplankton

113. Consider the following statements:

1. All domestic and municipal effluents be drained in water bodies after proper treatment
2. Only underground water be used for drinking purposes.
3. There must be strict regulation for industrial effluents, that must not discharge highly toxic water into water bodies.

Which of the above statement(s) is/are correct?

(a) only 1 (b) only 2
(c) 1 and 2 (d) 1 and 3

114. What is the maximum limit fixed for noise in most of the countries?

(a) 50 to 70 decibels (b) 75 to 85 decibels
(c) 80 to 90 decibels (d) 80 to 100 decibels

115. Which of the following is not a natural source of noise?

(a) Thunder (b) Volcanic eruption
(c) Rail noise (d) Sound of ocean waves

116. Which part of human body is highly affected by noise pollution?

(a) Lungs (b) Ear
(c) Nose (d) Heart

117. In which year Noise Pollution (Regulation and Control) Rules was introduced in India?

(a) 1985 (b) 1990
(c) 2000 (d) 1986

118. Which of the following is not a source of land pollution?

(a) Domestic and municipal waste

(b) Forest waste
(c) Industrial and mining waste
(d) Agricultural waste

119. In India 'NEERI' institute is located at:

(a) Nagpur
(b) Kolkata
(c) Mumbai
(d) Delhi

120. How the biological oxygen demand (BOD) gets affected with the increased pressure of organic matter in water?

(a) The oxygen demand increases.
(b) The oxygen demand decreases
(c) The oxygen demand remain unchanged
(d) None of the above

121. The best way to dispose plant waste is:

(a) burning (b) composting
(c) dumping (d) incineration

122. Which of the following statements is/are correct?

1. Soil pollution is responsible for less of fertility
2. Dumping of waste in oceans is a cause of imbalance of the marine ecosystem
3. Afforestation can be used as a measure to control land pollution

Select the correct answer using the codes given below:

(a) only 1 (b) only 2
(c) 1 and 2 (d) 1, 2 and 3

123. Which of the following would cause a rising water table?

(a) Increased precipitation
(b) Increased run-off
(c) Decreasing infiltration
(d) Groundwater depletion extending the exchange rate

124. Acidification of oceans is increasing. Why is this a cause of concern?

1. Growth and survival of calcareous phytoplankton will be adversely affected
2. The growth and survival of coral reefs will be adversely affected
3. The survival of some plants that have phytoplankton larva will be adversely affected.
4. The cloud seeding and formation of clouds will the adversely affected.

Which of the statement give below is/are correct:

(a) 1, 2 and 3 (b) only 2
(c) 1 and 3 (d) All of these

125. A pesticide, which is chlorinated hydrocarbon is sprayed on a food crop. The food chain is food crop-rat-snake-hawk. In this food chain, the highest concentration of the pesticide would accumulate in which of the following?

(a) Food crop (b) Rat
(c) Snake (d) Hawk

126. How can e-waste be treated in an environmentally sound way?

(a) Decontamination, dismantling and segregation
(b) Shredding

(c) Recovery of metals from e-waste
(d) All the above methods.

127. Fukushima is a associated with:

(a) Nuclear disaster (b) Volcanic eruption
(c) Cyclonic hazard (d) Air accident

128. Ecosystem management values most:

(a) ecological sustainability
(b) economic productivity
(c) recreational use of managed regions
(d) opportunities to harvest natural resources

129. Which one of the following is not a method to control land pollution?

(a) Sanitary landfill (b Incineration
(c) Disposal in water bodies (d) Composting

130. Which of the following waste material can not be recycled?

(a) Paper (b) Glass
(c) Tyre (d) Electronic goods

131. Consider the following statements:

1. Not-in-my-backyard
2. Not-in-anybody's-backyard
3. Not anywhere

Which of the above statement(s) is/are related with 'NIMBY' and 'NIABY'?

(a) only 1 (b) 1 and 2
(c) only 3 (d) 1, 2 and 3

132. Which of the following energy source is considered as eco-friendly?

(a) Solar energy (b) Coal
(c) Atomic energy (d) Petroleum

133. Which of the following state is suitable for the production of wind energy?

(a) Punjab (b) Bihar
(c) Jharkhand (d) Rajasthan

134. Which of the following State of India is least affected by flood?

(a) Bihar (b) Madhya Pradesh
(c) Uttar Pradesh (d) Assam

135. Consider the following:

1. Agricultural production can be increased only by the use of chemical pesticides and insecticides.
2. Excessive use of chemical pesticides and insecticides has resulted in degradation of environment.
3. Organic farming is an eco-friendly form of agriculture.

Which of the above statement(s) is/are suitable for sustainable development of agriculture?

(a) only 1 (b) 1 and 2
(c) 2 and 3 (d) 1 and 3

136. Which of the following country has suffered heavy destruction caused by tropical cyclones?

(a) Sri Lanka (b) Pakistan
(c) Bangladesh (d) Thailand

137. In which year was National Environment Tribunal Act introduced in India?

(a) 1986 (b) 1995

(c) 1998 (d) 2002

138. Which one of the following statements is not correct?

(a) The energy levels within a food chain are called trophic levels.

(b) In deep oceans, primary production is almost nil.

(c) The primary consumers are called autotrophs.

(d) The decomposers are called saprotrophs.

139. **Assertion (A):** The geothermal energy is considered clean, cost-effective and sustainable but there are some environmental concerns:

Reason (R): The fluids drawn from deep inside the earth carry a mixture of gases like carbon dioxide, ammonia, hydrogen sulphide and methane that are atmospheric pollutants.

Select the correct answer from the codes given below:

(a) Both **A** and **R** are true and **R** is the correct explanation of **A**

(b) Both **A** and **R** are false

(c) **A** is true but **R** is false

(d) **A** is false but **A** is true

Answers

1. b	2. a	3. c	4. c	5. d	6. c	7. d
8. a	9. b	10. a	11. b	12. c	13. b	14. c
15. a	16. b	17. c	18. a	19. d	20. d	21. a
22. c	23. b	24. b	25. a	26. c	27. d	28.a
29. d	30. b	31. d	32. a	33. c	34. b	35. d
36. b	37. a	38. b	39. a	40. c	41. a	42. b
43. c	44. c	45. b	46. b	47. a	48. a	49. a
50. c	51. b	52. c	53. d	54. c	55. d	56. c
57. a	58. d	59. a	60. c	61. b	62. d	63. c
64. c	65. d	66. b	67. d	68. b	69. d	70. b
71. a	72. c	73. c	74. b	75. a	76. b	77. c
78. a	79. c	80. c	81. d	82. b	83. a	84. d
85. b	86. c	87. b	88. c	89. b	90. b	91. a
92. c	93. d	94. b	95. c	96. b	97. c	98. b
99. a	100. d	101. b	102. a	103. d	104. c	105. c
106. b	107. b	108. d	109. c	110.a	111. c	112. b
113.d	114. b	115. c	116. b	117. c	118. b	119. a
120. a	121. b	122. c	123. a	124. a	125. d	126. d
127. a	128. a	129. c	130. d	131. b	132. a	133. d
134. b	135. c	136. c	137. b	138. c	139. a	

2 Biodiversity

Biodiversity or biological diversity is the foundation of life on the earth. It is crucial for the functioning of ecosystems which provide us with products and services without which we couldn't live. Biodiversity is extremely complex, dynamic and varied like other features of the earth. Its innumerable plants, animals and microbes physically and chemically unite the atmosphere, geosphere and hydrosphere into one environmental system which makes it possible for millions of species, including people, to exist. At the same time, no other feature of the earth has been so dramatically influenced by man's activities. By changing biodiversity, we strongly effect human well-being and the well-being of every other living creature. The increasing interest in biodiversity is a result of concern regarding species extinction, depletion of genetic diversity and disruptions to the atmosphere, water supplies, fisheries and forests.

Biodiversity

Biodiversity is a combination of two words 'biological' and 'diversity'. Biologists most often define biodiversity as the 'totality of genes, species and ecosystem of a region'. A few definitions of biodiversity are as given in Box 2.1.

In brief, biodiversity is that part of nature which includes the differences in genes among individuals of a species, the variety and richness of all the plant and animal species at different scales in space – locally, in a region, in the country and the world; and the types of ecosystems, within a defined area.

Box 2.1
Definitions of Biodiversity

1. 'Biodiversity is the full variety of the biosphere. All the species of living organisms, including plants, animals and microorganism. Biodiversity also includes the genetic variability within species and the biotic communities in which they interact'. (Joseph Kerski and Simon Ross, 2005)
2. 'A term describing the variety of species, both flora and/or fauna, contained within an ecosystem'. (John Small and Michael Withesick, 1995)
3. 'Biological diversity – an enormous cornucopi of wild and cultivated species, diverse in form and function, with beauty and usefulness beyond the wildest imagination'. (Illtis, 1998 quoted by Andrew Goudie, 2004)
4. United Nations Earth Summit (1992): Defined biological diversity as:
   'The variability among living organisms from all sources, including, "inter alia", terrestrial, marine and other aquatic ecosystems and the ecological complexes of which they are part: this includes diversity within species, between species and of ecosystems'.

The term 'biological diversity' was first used by wildlife scientist and conservationalist Ramond F. Dasmann in the 1968 lay book, *A Different Kind of Country*. The term was widely adopted only after more than a decade, when in the 1980s it came into common usage in science and environmental policy. Thomas Lovejoy, in the foreword of the book *'Conservation Biology'*, introduced the term to the scientific community. Until then the term 'natural diversity' was common, introduced by The Science Division of The Nature Conservancy in an important 1975 study 'The Preservation of Natural Diversity'. The term contracted from biodiversity may have been coined by W.G. Rosen in 1985, while planning National Forum on Biological Diversity. It first appeared in publication in 1988 when socio-biologist, E.O. Wilson used it as the title of the proceedings of that forum. Since this period, the term has achieved widespread use among biologists, environmentalists, political leaders, etc.

Types or Levels of Biodiversity

Biodiversity is usually described at three levels:

- Genetic diversity
- Species diversity
- Ecosystem diversity

In 2003, Anthony Compbell at Cardiff University, UK and the Darwin Centre Pembrokeshire, defined a fourth level: Molecular diversity.

Genetic Biodiversity

Each member of any animal or plant species differs widely from other individuals in its genetic makeup owing to the large number of combinations possible in the genes that give every individual its specific characteristics. Thus, for example, each human being is very different from all others. This genetic variability is essential for a healthy breeding population of species. If the number of breeding individuals is reduced, the dissimilarity of genetic makeup is reduced and inbreeding occurs. This leads to genetic anomalies and eventually, to the extinction of that particular species. It is the diversity in wild species from the 'gene pool' from which our crops and domestic animals have been developed over thousands of years. Today, the variety of nature's bounty is being further exploited by using wild relatives to create new varieties of more productive, disease-resistant crops and to breed superior domestic animals. Modern biotechnology also manipulates genes to develop better types of medicines and a variety of industrial products.

Species Biodiversity

The number of species of plants and animals that are present in a region constitutes its species diversity. The diversity is seen both in natural ecosystem and in agricultural ecosystem. Some areas are richer in species than others. For example, natural undisturbed tropical forests have much greater species' richness than monoculture plantations developed by the Forest Department for timber production. A natural forest ecosystem provides a large number of non-timber forest products (NTFPs) that local people depend on, such as fruit, fuelwood, fodder, fibre, gum, resin and medicines. Timber plantations do not provide the large variety of goods that are essential for local consumption. In the long-term, the economic sustainable returns from NTFPs is said to be greater than the returns from felling a forest for its timber. Thus, the commercial value of a natural forest, with all its species' richness, is much greater than a plantation. Modern intensive agricultural ecosystem has a relatively lower diversity of crops than traditional agro-pastoral farming systems, where multiple crops were planted. At present, conservation scientists have been able to identify and categorize about 1.8 million species on the earth. However, this is only a fraction of what really exists. Many new species are being identified, especially in the flowering plants and insects in areas that are rich in species diversity.

Ecosystems Biodiversity

An ecosystem is a set of life forms (plants, animals, microorganisms) interaction with one another and with non-living elements (soil, air, water, minerals, etc.). There are large varieties of different ecosystems on the earth, each having their own complement of distinctive interlinked species based on the differences in the habitat. Ecosystem diversity can be described for a specific geographical region, or a political entity such as a country, a state or a taluka. Distinctive ecosystems include landscapes like forests, grasslands, deserts, mountain, etc., as well as aquatic ecosystem like rivers, lakes and seas. Each region also has man-modified areas such as farmland of grazing pastures.

Two other forms of biodiversity (although not included in above-mentioned levels) that needs special mention are domesticated biodiversity and microorganism biodiversity.

Domesticated Biodiversity

When we think of biodiversity, we tend to think only of wild plants and animals. But there is also considerable diversity among domesticated plants and animals. Domesticated biodiversity may be the result of manipulation by humans, or of natural adaptations to different conditions over a period of time.

Since the dawn of agriculture, people in different parts of the world have developed different plant and animal varieties to meet certain needs and conditions. These include higher productivity, better taste, resistance to pests or disease, and the ability to withstand adverse conditions like floods, drought or frost. When humans first started cultivating cereals, they must have chosen those with characteristics that were suitable for easier cultivation and harvesting, for example, varieties of grains that had larger kernels, more rows of kernels on each stalk, or those whose seeds did not disperse as soon as they were ripe. This selection was done by storing and sowing seeds of few plants with the desired characteristics. Over a period of time, these developed into the ancestors of the cereal crops we know today.

Over time, this kind of adaptation to natural conditions or to human manipulation has led to large variety in the species of domesticated plants. The number of varieties of those grown in India is estimated to have been between 50,000 and 60,000. Other crops with high diversity include mango, 1,000 varieties; sorghum, 5,000 varieties and pepper, 500 varieties.

Microorganism Biodiversity

When we think of biodiversity, we rarely think of the most abundant organisms on earth: microorganisms or microbes. Microbes include bacteria, viruses, protozoa, yeast, fungus, etc., and form a vital part of life on earth. Bacteria are the oldest life forms on earth. Microbes play an important role in various biogeochemical cycles. They live in the digestive tracts of most animals (including humans and insects), where they break down the food and facilitate digestion. Microbes that live in the roots of leguminous plants transform atmospheric nitrogen and make it available to the plants (nitrogen fixation). The soil contains thousands of species of microbes which decompose dead organic matter and help maintain soil structure. Just a teaspoonful of soil contains billions of these microscopic living organisms.

Mapping Species Biodiversity

Brits, Paul Williams, Decair Right and Chris Humpreger developed a software to map biodiversity of the world. They have identified three types of diversities for mapping. These are:

Alpha Diversity: The total number of species found in a particular region. Thus a comparative analysis of the regional pattern of biodiversity can be done.

Beta Diversity: It is structural diversity found in species in particular regions.

Gama Diversity: It shows rate of change of species with the expanding geographical region.

Importance of Biodiversity

Biodiversity provides a variety of environmental services through its species and ecosystem that are essential at the global, regional and local levels. The production of oxygen, reduction of carbon dioxide, maintaining the water cycle, and protecting soil are some important services. The world now acknowledges that the loss of biodiversity contributes to global climatic changes. Forests are the main mechanism for the conversion of carbon dioxide into carbon and oxygen. The loss of forest cover, coupled with the increasing release of carbon dioxide and other gases through industrialization contributes to the 'greenhouse effect'. Global warming is melting ice caps, resulting in a rise in the sea level, which will eventually submerge the low lying areas in the world. It is also causing major atmospheric changes, leading to increased temperatures, serious drought in some areas and unexpected floods in other areas.

Biological diversity is also essential for preserving ecological processes, such as fixing and recycling of nutrients, soil formation, circulation and cleansing of air and water, global life support (plants absorb CO_2, give out O_2), maintaining the water balance within ecosystems, watershed protection, maintaining stream and river flows throughout the year along with erosion control and local flood reduction.

Food, clothing, housing, energy and medicines are all resources that are directly or indirectly linked to the biological variety present in the biosphere. This is most obvious in the case of tribal communities who directly gather resources from the forest, or fisherfolk who catch fish in marine or fresh water ecosystems. For others, such as agricultural communities, biodiversity is used to grow their crops to suit the environment. Urban communities generally use the greatest of goods and services, which are all indirectly drawn from natural ecosystems.

Broadly speaking, biodiversity is having immense importance in man's life and in socio-economic development. The following facts illustrate the importance of biodiversity in various fields.

Consumptive Importance

The biodiversity contained in the ecosystem provides forest dwellers with all their daily needs, food, building material, fodder, medicines and variety of other products. They know the qualities and different uses of wood from different species of trees, and collect a large number of local fruits, roots and plant material that they use as food, construction material or medicines. Fisherfolk are completely dependent on fish and know where and how to catch fish and other edible aquatic animals and plants. Perhaps the most important value of biodiversity, particularly in a country like India, is that it meets the basic survival needs of a vast number of people. Even today a large number of traditional communities depend, wholly or partially on the surrounding natural resources for their daily needs of food, shelter, clothing, household goods, medicines, etc.

Productive Use Value

This category comprises of marketable goods. The biotechnologist uses bio-rich areas to 'prospect' and search for potential genetic properties in plants or animals that can be used to develop better varieties of crops for use in farming and plantation programmes or to develop better livestock. To the pharmacist, biological diversity is the raw material from which new drugs can be identified from plant or animal products. To the industrialist, biodiversity is a rich storehouse from which to develop new products. For the agricultural scientist, the biodiversity in the wild relatives of crop plants is the basis for developing better crops.

Genetic diversity enables scientist and farmers to selectively develop better crops and domestic animals through careful breeding programmes. Earlier, this was done by selecting or pollinating crops artificially to get a more productive or disease-resistant strain. Today it is increasingly being done by genetic engineering – selecting genes from one plant and introducing them into another. New crop varieties (cultivars) are being developed using the genetic material found in wild relatives of crop plant through biotechnology.

The preservation of biodiversity has now become essential for industrial growth and economic development. A variety of industries, like pharmaceuticals are highly dependent on identifying compound of great economic value from the wide variety of wild species of plants located in undisturbed natural forests. This is called 'biological prospecting'. Up to 80 per cent of the people in developing countries depend for primary healthcare on traditional medicines, most of which is derived from plants, and some from animal and mineral sources. Even several modern drugs have been derived from plant source. Some of such drugs, their source and use is given in Box 2.2.

Box 2.2
Commonly Used Modern Drugs and their Plant Sources

Drug	*Plant Source*	*Use*
Atropine	Belladonna	Anticholinergic, reduces intestinal pain in diarrhoea
Bromelain	Pineapple	Controls tissue inflammation due to infection
Caffeine	Tea, coffee	Stimulant of the central nervous systems
Camphor	Camphor tree	Ruberfacient; increases local blood supply
Cocaine	Cocoa	Analgesic and local anaesthetic; reduces pain and prevents pain during surgery
Codeine	Opium poppy	Analgesic; reduces pain
Morphine	Opium poppy	Analgesic, controls pain
Colchicines	Autumn crocus	Anticancer agent
Digitoxin	Common foxglove	Cardiac stimulant used in heart disease
Diosgenin	Wild yams	Source of female contraceptive; prevents pregnancy
L-dopa	Velvet bean	Control Parkinson's Disease, which leads to jerky movements of the hands
Ergotamine	Smut-of-rye ergot	Control of haemorrhage and migraine headaches
Glaziovine	Ocotea glaziovii	Antidepressant agent
Gossypol	Cotton	Male contraceptive

Cont'd...

...Cont'd

Indicine N-oxide	Heliotropium indicum	Anticancer agent
Menthol	Mint	Ruberfacient; increases local blood supply and reduces pain on local application
Monocrotaline	Cotolaria sessiliflora	Anticancer agent
Papain	Papaya	Dissolves excess protein and mucus, during digestion
Penicillin	Penicilluim fungi	General antibiotic, kills bacteria and controls infection by various microorganisms
Quinnine	Yellow cinchona	Antimalarial
Reserpine	Indian snakeroot	Reduces high blood pressure
Scopolamine	Thorn apple	Sedative
Taxol	Pacific yew	Anticancer (ovarian)
Vinblastine	Rosy periwinkle	Anticancer agent; controls cancer in children
Vincristine	Vinca rosea (sadaphali)	—

Ethical, Religious and Cultural Importance

Ethical values related to biodiversity conservation are based on the importance of protecting all forms of life. Most religious and secular creeds believe that all forms of life have the right to exist on the earth. Man is only a small part of the earth's great family of species; plants and animals have an equal right to live and exist on our planet. We do not know if life as we know it exists anywhere else in the universe. Do we have the right to destroy life-forms or do we have a duty to protect them?

Apart from the economic importance of conserving biodiversity, there are several cultural, moral and ethical values, which are associated with the sanctity of all forms of life. Indian civilization has, over several generations, preserved nature through local traditions. This has been an important part of the ancient philosophy of many of our cultures. We have in our country a large number of sacred groves or *deorais* preserved by tribal people in several states. These sacred groves around ancient sacred sites and temples act as gene banks for wild plants.

In India, many plants and animals having ritual significance are associated with religious, spiritual and other cultural uses. Among the auspicious flowers offered in temples are hibiscus, offered to goddess Kali, and *datura* flower offered to Shiva. Various plant and animal species are considered sacred on account of their association with different deities. Some animal species are believed to be the *vahanas* or vehicles of the deities and are hence venerated. Important among these are the bull of Shiva, the rat of Ganesha and the lion of Durga.

Aesthetic Value

The appreciation of the presence of biodiversity for its inherent value and beauty, as well as for the contribution it makes to our knowledge – our aesthetics, imagination and creativity – are other reasons to preserve it. Quite apart from killing wildlife for food, it is important as a tourist attraction. Watch a spider weave its complex web. Observe a fish feeding, it is magnificent and fascinating. Each species and ecosystem adds to the richness and beauty of life on earth. Perhaps no artificial medium can match the sheer joy of watching a sunset over an ocean, the sight of a leaping deer, the sound of a singing bird, or the smell of wet earth after the first rains. A natural ecosystem, once destroyed, is impossible to recreate.

Ecological Services

Biodiversity supports many ecosystem services that are often not readily visible and it plays a part in regulating the chemistry of our atmosphere and water supply. Biodiversity is directly involved in water purification, recycling nutrients and providing fertile soil.

Species evolve to fill particular niches (roles) in an ecosystem or habitat. Many species also depend on each other in intricate ways for survival. Destroying one species can lead to further extinctions or changes. Specific life forms present in a particular habitat help to create conditions for other life forms to live in that environment. For example, a single tree provides not only its products, which may have economic value, but it is also a habitat for innumerable living things. In addition, it also plays a vital role in conserving soil and water and helping to keep the air clean. Mangroves and coral reefs, apart from their normal ecosystem service have a role in preventing erosion, and play a critical role in protecting offshore life during hurricanes and storms.

Apart from above-mentioned values of biodiversity, it also has relevance to human health. This issue is closely linked with the issue of climate change, and many of the anticipated health risks of climate change are associated with changes in biodiversity. Some of the health issues influenced by biodiversity include dietary health, nutrition security, infectious disease, medical science and medicinal resources, social and psychological health. Biodiversity is also known to have an important role in reducing disaster risk and in post-disaster relief and recovery efforts.

Number of Species

How much biodiversity of species is there? Estimates range widely, and the fact is, no one knows. Today, the only two certainties are that 1.75 million species have been described and many more than that exist. Most people are completely unaware of the great diversity of species within any given taxonomic category. Groups, especially rich in species are the flowering plants (270,000 species) and the insects (950,000 species), but even less diverse groups, such as birds or ferns, are rich with species that are unknown to most people.

Table 2.1 *Estimated Species on Earth and Oceans*

Species	*Earth*			*Ocean*		
	Catalogued	*Predicted*	*± SE*	*Catalogued*	*Predicted*	*± SE*
Eukaryotes						
Animalla	953,434	7,770,000	958,000	171,052	2,150,000	145,000
Chromista	13,033	27,500	30,500	4,859	7,400	9,640
Fungi	43,271	611,000	297,000	1,097	5,320	11,100
Plantae	215,644	298,000	8,200	8,600	16,600	9,130
Protozoa	8,118	36,400	6,690	8,118	36,400	6,690
Total	1,233,500	8,740,000	1,300,000	193,756	2,210,000	182,000
Prokaryotes						
Archaea	502	455	160	1	1	0
Bacteria	10,358	9,680	3,470	652	1,320	436
Total	10,860	10,100	3,630	653	1,321	436
Grand Total	1,244,360	8,750,000	1,300,000	194,409	2,210,000	182,000

Source: Mora, Camilo; Tittenson, Derek P., Adl, Sina' Simpson, Alastair, G.B.; Worm, Boris, Mace Georginam (23 August 2011), 'How many species are there on Earth and in the Ocean?'.

According to Mora and colleagues, the total number of terrestrial species is estimated to be around 8.7 million, while the number of oceanic species is much lower, estimated at 2.2 million.

According to the Global Taxonomy, Initiative and the European Distributed Institute of Taxonomy the total number of species for some phyla may be much higher than what is shown in table:

- 10–30 million insect; (of some 0.9 million we know today)
- 5–10 million bacteria;
- 1.5 million fungi; (of some 0.075 million we know today)
- 1 million mites.

Taxonomists are aware that their work in finding and describing new species is incomplete. Whatever the number, the plant's biodiversity represents an amazing and diverse storehouse of biological wealth.

Distribution of Biodiversity

Biodiversity is not evenly distributed; rather it varies greatly across the globe as well as within regions. Among other factors, the diversity of all living things (biota) depends on temperature, precipitation, altitude, soil, geography and the presence of other species. The study of the spatial distribution of organisms, species and ecosystems, is the science of biogeography.

Diversity consistently measures higher in the tropics and in other localized regions such as the Cape Floristic Region and lower in the polar regions generally. Rainforest that have had wet climates for a long time, such as Yasuni National Park in Ecuador, have particularly high biodiversity.

Terrestrial biodiversity is up to 25 times greater than ocean biodiversity. Although a recently discovered method put the total number of species on the earth at 8.7 million of which 2.1 million were estimated to live in the ocean.

Generally, there is an increase in biodiversity from the poles to the tropics. Thus, location at lower latitudes have more species than localities at higher latitudes. This is often referred to as the latitudinal gradient in species diversity. Several ecological mechanisms may contribute to the gradient, but the ultimate factor behind many of them is the greater mean temperature at the equator compared to the poles.

Even though terrestrial biodiversity declines from the equator to the poles, some studies claim that this characteristic is unverified in aquatic ecosystem, especially in marine ecosystem. The latitudinal distribution of parasites does not follow this rule.

Tropical regions of the world are having highest biodiversity. Even in tropics three species regions are having rich biodiversity. These regions are: (i) tropical rainforests, (ii) coral reefs and (iii) wetlands.

Tropical rainforests are unique and richest in terms of number of species. Covering just 7 per cent of the earth's surface, they are home to nearly half of all known species. According to the US National Academy of Sciences, a typical patch of rainforest about 10 sq km contains as many as 1,500 species of flowering plants, 750 tree species, 400 bird species, 150 butterfly species, 100 reptile species, 60 amphibian species, an unknown number of insects species. Biologist E.O. Wlison identified 43 species of ants on a single tree in a Peruvian rainforest. Unfortunately, the same tropical forests that hold such high biodiversity are also experiencing the highest rate of deforestation.

Coral reefs are also rich in biodiversity because of the abundance and diversity of life found there. Australian Great Barrier Reef is the world's largest coral reef ecosystem, containing more than 3,000 animal species. The richest, most diverse coral reefs in the world are found in the Indo-West Pacific. These reefs contain more that 2,000 species of fish, 5,000 species of molluscs, 700 species of coral and countless species of crabs, sea urchins brittle stars, sea cucumbers and worms. Coral reefs provide habitats for many organisms that cannot live elsewhere. The coral reefs of Australia, the Philippines, Indonesia, Malaysia, Japan, Palau, Maldives, Tanzania, Seycholles, Belize and Florida are very rich in biodiversity, but at the same time are the worst hit areas by variety of human activities.

Wetlands are transitional zones between land and water and are among the most biologically interesting and productive habitats. There are two types of wetlands: *inland* (fresh water ecosystems including marshes, swamps, riverine wetland and bogs) and *coastal* (either fresh or salt water mangroves). Fresh water marshes are important habitats for wildlife and migratory birds, while swamp are dominated by woody vegetation, reptiles, amphibians and fish. Tidal salt marshes occur along the coastlines of mid-latitudes and are dominated by mangroves in subtropics and tropics. Mangrove swamp is a dominant type of coastal wetland. These are forested wetlands and cover approximately 16 million hectare worldwide, with largest concentration in tropical Asia. Both coastal and inland wetlands are the reservoirs of biodiversity.

The fact is, that most of the world's bio-rich nations are in the South, in other words, they are the 'developing' nations. In contrast, the majority of the countries capable of exploiting biodiversity are northern nations, in the economically 'developed' world.

India as a Mega-Diversity Nation

India is one of the world's 12 mega-diversity countries. India contains 8.1 per cent of the world biodiversity on 2.4 per cent of the earth's surface. It is estimated that 47,000 wild species of plants and over 89,450 wild species of animals occur in our country. India has a tremendous range of ecosystem, species and genetic diversity. The location of the Indian subcontinent at the confluence of three biogeographic realms has led to the presence of elements of African, European, Chinese and Indo-Malayan characteristics in the flora and fauna. The diversity is also due to the fact that India has almost every major type of habitat and climatic condition – from alpine height to coasts and plains; from areas of the heaviest rainfall to dry deserts.

India is a bio-rich country having great variety of plants and animals, many of which are not found elsewhere.

Table 2.2 *Biodiversity of India and its World Ranking*

	India's world ranking	*Number of species in India*
Mammals	8th	350
Birds	8th	1,200
Reptiles	5th	453
Amphibian	15th	182
Angiosperms	15th–20th	14,500

India has 350 different mammals (rated the eighth highest in the world), 1,200 species of birds (eighth in the world), 453 species of reptiles (fifth in the world), 182 species of amphibian (fifteenth in the world) and 14,500 plant species, of which most are angiosperms (fifteenth in the world). These include, especially, high species diversity of ferns (1,022 species) and orchids (1,082 species). India has 50,000 known species of insects, including 13,000 butterflies and moths. It is estimated that the number of unknown species could be several times higher.

It is estimated that 18 per cent of Indian plants are endemic to the country and found nowhere else in the world. Among the plant species, the flowering plants have much higher degree of endemism, a third of these are not found elsewhere in the world. Among amphibians found in India, 62 per cent are unique to this country. Among lizards, of the 153 species recorded, 50 per cent are endemic. High endemism has also been recorded for various groups of insects, marine worms, centipedes, mayflies and freshwater sponges. Apart from high biodiversity in Indian wild plant and animals, there is also a great diversity of cultivated crops and breeds of domestic livestock.

India is also considered one of the world's eight centres of origin of cultivated plants. It has 51 species of cereals and millets, 104 species of fruits, 27 species of spices and condiments, 55 species of vegetables and pulses, 24 species of fibre crops, 12 species of oilseeds, and various wild strains of tea, coffee, tobacco and sugar cane. India also has significant indigenous livestock diversity, with 27 breeds of cattle, 40 breeds of sheep and 22 breeds of goats. For example, India's eight breeds of buffaloes represent the entire range of genetic diversity of buffaloes in the world. Many of these today have died out or are dying out, due to our misguided adoption of all 'foreign' things. Jerseys and Holsteins have largely replaced Indian cattle like Brahma bulls; and high-yielding cultivars have eaten away into centuries old landraces of crops.

India can be conveniently divided into ten major regions, based on the geography, climate and pattern of vegetation seen and the communities of mammals, birds, reptiles, amphibians, insects and other invertebrates that live in them. Each of these regions contains a variety of ecosystems such as forests, grasslands, lakes, wetlands, mountains and hills, which have specific plant and animal species.

India's biogeographic zones are as follows:

1. The cold mountainous snow covered Trans-Himalayan region of Ladakh.

2. The Himalayan ranges and valleys of Kashmir, Himachal Pradesh, Uttarakhand, Assam and other north-eastern States.
3. The *Terai,* the lowland where the Himalayan rivers flow into the plains.
4. The Gangetic and Brahmaputra plains.
5. The Thar Desert of Rajasthan.
6. The semi-arid grassland region of the Deccan plateau, Gujarat, Maharashtra, Andhra Pradesh, Karnataka and Tamil Nadu.
7. The north-eastern States of India.
8. The Western Ghats in Maharashtra, Karnataka and Kerala.
9. The Andaman and Nicobar Islands.
10. The long western and eastern coastal belt with sandy beaches, forest and mangroves.

Decline of Biodiversity

The extinction of several species or loss of biodiversity is a major problem of the international world. Species are disappearing rapidly and there is growing consensus within the international community that a system should be put in place to slow or halt the process of extinction. All the species are the integral part of ecosystem and extinction of some species diminishes the well-being of the remaining species, including human beings. The concern about biological diversity is the shrinking genetic pool. It is estimated that tropical forests contain at least 50 per cent and perhaps, 90 per cent of the world's species. According to an estimate, 20 to 75 species are becoming extinct each day because of deforestation in the tropics.

Species do not persist indefinitely. Indeed, it was the realization that fossils are the remains or traces of organism no longer to be found alive that led to the development of our present understanding of evolutionary process. Estimating the number of species existing at present in proportion of all the species that have ever lived is difficult, but the plants and animals alive today probably constitute around 2–4 per cent of all the species that have lived in the last 600 million years.

Extinctions have not been spread evenly. There is a general background rate at which species go extinct, but there have also been at least five mass extinctions in which 65–85 per cent(and at the end of the Permian Period, around 225 million year ago, 95 per cent or more) of all species disappeared. Background extinctions have little or no effect on biodiversity, because their rate is matched or exceeded by that of speciation, in which one species divides into two and thus increases the number of species. Nor do mass extinctions produce more than a temporary reduction in the number of species, because historically they have been followed by the rapid adaptive radiation of surviving species into the vacated niches. The extinction of the dinosaurs, for example, was followed by rapid speciation among mammals. Today the number of land-dwelling species alive is about twice the average of the last 450 million years.

Man has begun to overuse or misuse most of these natural ecosystems. Due to this unsustainable resource use, once productive forest and grasslands have been turned into desert and wastelands have increased all over the world. Mangroves have been cleared for fuelwood and prawn farming, which has led to a decrease in the habitat essential for the breeding of marine fish. Wetlands have been drained to increase agricultural land. These changes will have grave economic implication in the longer term.

As mentioned earlier, there are about 1.8 million species of plants and animals known to Science in the world at present. Unfortunately, at the present rate of extinction, about 25 per cent of the world's species will undergo extinction fairly rapidly. This may occur at the rate of 10–20,000 species per year, at a 1000–10,000 faster rate than the expected natural rate! Human actions could well exterminate 25 per cent of the world's species within the next twenty or thirty years. Much of this mega-extinction spasm is related to human population growth, industrialization and changes in land use patterns. A major part of these extinctions will, of course, occur in bio-rich areas such as tropical forests, wetlands, and coral reefs. The loss of wild habitats, due to rapid human population growth and short-term economic development, is the contributor to the rapid global destruction of biodiversity.

India's rich biological diversity is rapidly eroding. At least 10 per cent of its recorded flora, and possibly a large fraction of its wild fauna, is threatened. Many may be on the verge of extinction. In the last few decades, India has lost at least 50 per cent of its forest; polluted over 70 per cent of its waterbodies; built, cultivated or otherwise encroached upon its grasslands and degraded many coastal areas.

The cheetah and the pin-headed duck are amongst the conspicuous species that have become extinct. More than 150 of the known species of medicinal plants in India have already become extinct due to unsustainable methods of harvesting. Of the species of flora and fauna that remain, over 10 per cent of flowering plants, 21 per cent of mammals and 5 per cent of birds are believed to be under threat. India's domesticated biodiversity is also under threat. Hundreds of crop varieties have disappeared, and even their genes have not been preserved.

The World Conservation Monitoring Centre (WCMC) has evaluated and described the threat to about 88,000 plants and 2,000 animal species in its series of Red Data Books. Threatened species include those which are endangered, vulnerable and rare in International Union for Conservation of Nature and Natural Resources (IUCN) categories. The great majority of species on these lists of Red Data Book are plants. However, there are also species of fish (343), amphibians (50), reptiles (170), invertebrate (1,355), birds (1,037), and mammals (497). India contains 172 species of animals considered globally threatened by IUCN.

Hot Spots

A biodiversity hotspot is a region with a high level of endemic species that are under threat from humans. The term 'hotspot' was introduced in 1988 by Dr. Sabina Virk. While hotspots are spread all over the world, the majority are forest areas and most are located in the tropics.

Brazil's Atlantic Forest is considered one such hotspot, containing roughly 20,000 plant species, 1,350 vertebrates, and millions of insects, about half of which occur nowhere else. The island of Madagascar, particularly the unique Madagascar dry deciduous forest and lowland rainforest, possess a high ratio of endemism. Since it separated from mainland Africa, 65 million years ago, many species and ecosystems have evolved independently. Indonesia's 17,000 islands cover 735,355 square miles (1,904,560 km^2) that contain 10 per cent of the world's flowering plants, 12 per cent of mammals and 17 per cent of reptiles, amphibians and birds – along with nearly 240 million people. Many regions of high biodiversity and/or endemism arise from specialized habitats which require unusual adaptations, for example alpine environment in high mountains of Northern European peat bogs.

In the world, 25 regions have been identified as biodiversity 'hot spots'. These are as follows:

Box 2.3
Hot Spots of the World

1.	Tropical Andes	13.	Succulent Karoo
2.	Mesomaerica	14.	Mediterranean Basin
3.	Caribbean	15.	Caucasus
4.	Brazil's Atlantic Forest	16.	Sundaland
5.	Choco/Darien/Western Ecuador	17.	Wallacea
6.	Brazil's Cerrado	18.	Philippines
7.	Central Chile	19.	Indo-Burma
8.	California Floristic Province	20.	South-Central China
9.	Madagascar	21.	Western Ghats/Sri Lanka
10.	Eastern Arc and Central Forest of Tanzania/Kenya	22.	SW Australia
		23.	New Caledonia
11.	West African Forest	24.	New Zealand
12.	Cape Floristic Province	25.	Polynesia/Micronesia

Major 'hot spots' are located in tropical and sub-tropical zones as shown in Figure 2.1.

Figure 2.1 *Major 'Hot Spots' in Tropical and Sub-tropical Zones*

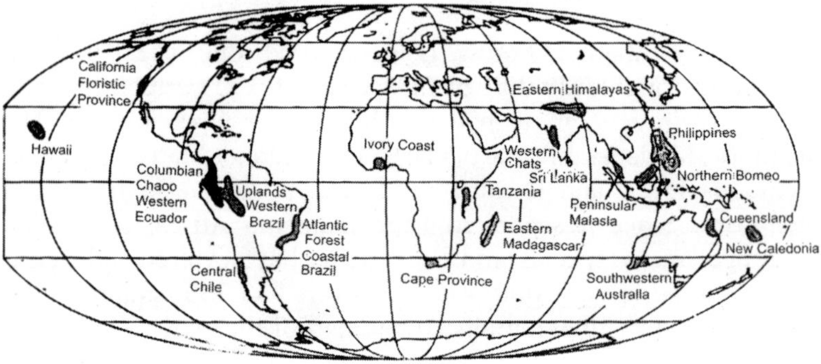

Causes of the Decline or Loss of Biodiversity

The loss of biodiversity throughout the world has become a common phenomenon. The known causes of extinctions are the destruction of habitats, introduction of foreign species and hunting, etc. Jared Diamond describes an 'Evil Quartet' of habitat destruction, overkill, introduced species, and secondary extinctions. While Edward O. Wilson prefers the acronyn HIPPO, standing for habitat destruction, invasive species, pollution, human over-population and over-harvesting as main causes. Figure 2.2

Figure 2.2 *Causes and Mechanics of the Loss of Biodiversity*

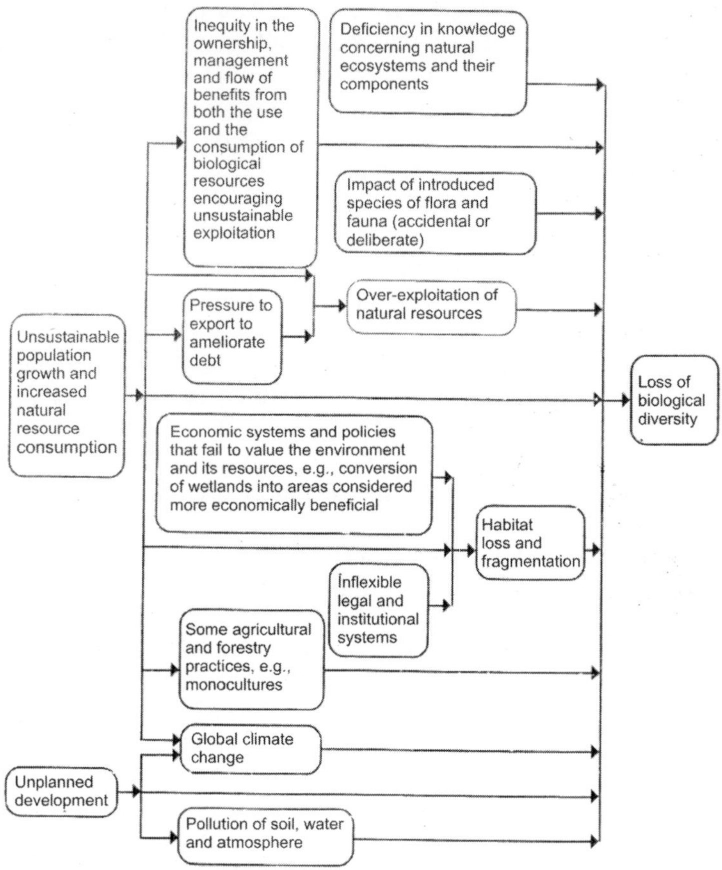

depicts the cause of the loss of biodiversity as identified in Global Biodiversity Strategy, WRI, IUCN and UNEP.

Habitat Destruction

One of the main causes of the loss of biodiversity is the physical alternation or destruction of habitats through the processes of conversion, fragmentation, and simplification. Habitat destruction has already been responsible for 36 per cent of the known extinctions.

Conversion: Natural areas are converted to farms, housing subdivision, shopping malls, marinas, and industrial centres. When a forest is cleared, for example, it is not just the trees that are destroyed – every other plant or animal that occupies the destroyed ecosystem, either permanently or temporarily (e.g. migrating birds), also suffers. The idea that the wildlife simply will move 'next door' and continue to live in an undisturbed section is erroneous. Any loss of natural habitat can result in only one thing – a proportional reduction in all populations that require that habitat.

Fragmentation: Natural landscapes generally have large patches of habitat that are well connected to other similar patches. Human-dominated landscapes, however, consist of a mosaic of different land uses, resulting in small, often geometrically configured patches that frequently contrast highly with neighbouring patches. Whether it be plowed fields of house lots, the tendency is to shrink the size of forested parcels. For the continued survival of any natural population, the number of individuals must never fall below a *critical number*, and that requires a certain minimum area that must be large enough to compensate for years of adverse weather. For many species, more area will be required during a dry year than during a normal year. If development reduces the habitat to a point where it cannot support the critical number during an adverse year, the entire population will perish. Similarly, development (such as a highway) that fragments a territory and prevents migration between the two fragments will cause a population to perish if neither area can adequately support the critical number. Also, reducing the size of a habitat creates a greater proportion of edges, a situation that favours some species, but may be detrimental to others.

Simplification: Human use of habitats often simplifies them. Removing fallen logs and dead trees from woodlands for firewood, for example, diminishes an important microhabitat on which several species depend. When a forest is managed for the production of a few or one species of trees, tree diversity declines, and with it so does the diversity of the cluster of plant and animal species that depend on the less favoured trees. Streams are sometimes 'channelized' – their beds are cleared of fallen trees and riffles, and sometimes the stream is straightened out by dredging. Such alterations inevitably reduce the diversity of fish and invertebrates that live in the stream.

Introduced and Invasive Species or Exotic Species

The introduction of non-native species deliberately or accidentally, has been a major threat to biodiversity worldwide. An exotic species is a species introduced into an area from somewhere else, often a different continent. Because the species is not a native of the new area, it is frequently unsuccessful in establishing a viable population and quietly disappears. This is the fate of many birds, reptiles, and fish that escape or are deliberately released from their native habitats. Occasionally, however, an introduced species finds the new environment very much to its liking and can become an *invasive species*, thriving, spreading out, and perhaps eliminating native species by predation or competition for space of food.

The introduced animals and plants pose a threat to the local species of fauna and flora. For example, the spotted deer (*Axis axis)* was introduced by the British in the Andaman and Nicobar Islands. The deer have now proliferated on these islands as they have no predator except crocodiles and humans; their large numbers seem to be affecting forest regeneration as they overexploit certain forest species for food. The deer are also causing crop damage in the fields of the island's settlers. Exotic plants, such as *Lantana Camara*, originally introduced as an ornamental plant from Brazil, are spreading rapidly in our forest at the expense of local species.

Exotics such as eucalyptus are replacing native tree species because they are fast growing and commercially valuable. It is estimated that 18 per cent of Indian flora comprises invasive aliens, of which about 55 per cent are American, 10 per cent Asian, 20 per cent Asian and Malaysian, and 15 per cent European and Central Asian species. At present, several countries have already imported so many exotic species, particularly agricultural and ornamental plants, that their own indigenous fauna/flora may be out numbered.

Overexploitation

Overexploitation occurs when a resource is consumed at an unsustainable rate. This occurs on land, in the form of overhunting, excessive logging, poor soil conservation in agriculture and the illegal wildlife trade. Joe Walston, director of the Wildlife Conservation Society's Asian programmes, called the latter the 'single largest threat' to biodiversity in Asia. The international trade of endangered species is second in size only to drug trafficking. About 25 per cent of world fisheries are now overfished to the point where their current biomass is less than the level that maximizes their sustainable yield.

Many plant and animal species have been overexploited by humans, sometimes to the point of extinction. Many species such as tigers and elephants are killed or poached for their skin, tusks, claws, etc. which have high commercial value. Others, such as several snake and bird species are caught and smuggled out for their curiosity value for collectors and as pets. Marine fauna are under great threat from overexploitation, largely as a result of mechanized fishing and increasing international fishing operations in Indian waters.

Trade in exotics is another form of overuse. Trafficking in wildlife and in products derived from wild species, although illegal, is a main reason for the loss of several species.

It flourishes because some consumers are willing to pay exorbitant prices for such things as furniture made from tropical hardwoods (like teak), exotic pets, furs from wild animals, traditional medicines from animal parts, and innumerable other 'luxuries' including polar bear, rugs, baskets made from elephant and rhinoceros feet, ivory-handled knives, and reptile-skin shoes and handbags. Tiger and rhinoceros populations in the wild have declined drastically in the last two decades because of the widespread, but unfounded belief in Far Eastern countries that parts from these animals have medicinal or aphrodisiac properties.

Poor management represents another form of overuse leading to a loss of biodiversity. Forests and woodlands are overcut for firewood, grasslands are overgrazed, game species are overhunted, fisheries are overexploited, and croplands are overcultivated.

Pollution

Another major factor that decrease biodiversity is pollution, which can directly kill many kinds of plants and animals, seriously reducing their populations. Soil, water and air pollution affects the functioning of ecosystems and may reduce or eliminate sensitive species. Several studies carried out in India have clearly traced the effects of pesticide pollution on populations. Every oil spill kills sea birds and, often, sea mammals, sometimes by thousands. Human wastes can spread pathogenic microorganisms to wild species, a threat called 'pathogen pollution'.

Climate Change

Global climate change brought on by anthropogenic greenhouse gas emissions, is already having an impact on many species. In the polar north, polar bears are in decline because the sea ice they depend on for hunting seals is melting at a rapid rate (up to 9% per decade). In the polar south, many penguin species are in decline as waters undergo unusual warming that disrupts the fish and invertebrate species that the penguins depend on. Rising land temperatures are already encouraging the spread of vector-borne diseases (diseases carried by ticks, files, mosquitoes, etc.) such as Lyme disease, malaria, dengue, and West Nile disease. Some of these are pathogens for both human and wildlife.

In the coming years, climate change could also affect global biodiversity. There are several hypotheses about this. Different species of plant and animals may respond in different ways to an increase in global temperature. Many species which cannot adjust to warmer temperatures could become extinct. Species with specialized niches or rare species will be at greatest risk being most sensitive to any atmospheric change. A change in the climate may also result in a change in the characteristics of habitats, thereby affecting the species within those habitats. Some habitats such as islands and coastal systems, which are at risk of flooding and submergence due to rising sea level, could suffer particularly high losses of biodiversity.

Population Growth

Past losses of biodiversity can be attributed to the expansion of the human population over the globe. Continuing human population growth will further alter natural ecosystems, resulting in the inevitable loss of more wild species and additional decline in populations. The losses will be greatest in the developing world, where biodiversity is greatest and human population growth is highest. Africa and Asia have lost almost two-thirds of their original natural habitat. People's desire for a better way of life, the desperate poverty of rural populations and the global market for timber and other natural resources are powerful forces that will continue to narrow down biological wealth on those continuers.

From 1950 to 2011, world population increased from 2.5 billion to 7 billion and is forecast to reach a plateau of more than 9 billion during the 21st century. Sir David King, former chief scientific adviser to UK government, told a parliamentary inquiry: 'it is self-evident that the massive growth in the human population through the 20th century has had more impact on biodiversity than any other single factor'.

Other causes of the loss of biodiversity include hybridization, genetic pollution, loss of traditional knowledge, ineffective legal system, changing agricultural and forestry practices, growing demands, international trade and management system based on exploitation of natural resources, etc.

Conservation of Biodiversity

The overall discussion on biodiversity reveals that this invaluable heritage is being destroyed at an alarming rate. All the nations of the world have shown their concern regarding conservation of biodiversity and several measures have also been taken in this direction. Prior to describing the measures of conservation, a special mention should be made of International Union for Conservation of Nature and Natural Resources (IUCN), now known as World Conservation Union (WCU), established in 1988. It has prepared and revised from time to time a list of the eatened plants and animals. Such lists are called Red Data Book or Red Data Lists. The Red List of threatened species released by IUCN includes the following categories:

Extinct Species (EX) are those species of plants and animal which have been completely eliminated.

Extinct in Wild (EW) are those species which have been eliminated but still exists in wild areas.

Critically Endangered (CR) are those species which have lost their 80 per cent population.

Endangered Species (EN) are those which have lost their 70 per cent of population within the last 10 years.

Vulnerable Species (VU) are those which are vulnerable to extinction and have lost 50 per cent of population.

Near Threatened Species (NT) are those species which are above the critically endangered and vulnerable species that has had high loss of population.

Least Concern (LC) are those species which have abundant population and require least attention.

The conservation measures of biodiversity can be divided into two categories:

1. *In situ* conservation.
2. *Ex situ* conservation.

In situ Conservation

In situ conservation, as defined by the UNEP (United Nations Environment Programme), is the conservation of ecosystems and natural habitats and the maintenance and recovery of viable populations of species in their natural surrounding. This is an effective conservation strategy since it ensures the maintenance of ecosystem and species in their natural conditions.

Biodiversity at all its levels, as genetic species and as intact ecosystem, can be best preserved *in situ* by setting aside an adequate representation of wildness as Protected Areas. These should consist of a network of National Parks and Wildlife Sanctuaries with each distinctive ecosystem included in their network. Such a network would preserve the total diversity of life of a region.

India has taken several steps towards *in situ* conservation of biodiversity.

India's major attempt to conserve *in situ* wild biodiversity has been through a network of protected areas throughout the country. Two kinds of protected areas are recognized by the legal system in India; national parks and wildlife sanctuaries. National parks are highly protected by law. No human habitation, private landholding or traditional human activity such as firewood collection or grazing is allowed within the park. Sanctuaries are also protected, but certain types of activities such as collection of firewood and grazing of cattle are permitted within these areas.

The first such initiative was the establishment of the Corbett National Park in 1936. India has 99 national parks and 513 wildlife sanctuaries occupying 156,000 sq km, or over 5 per cent of the country's area. These protected areas have helped in conserving habitats and their biodiversity.

Several special projects have also been launched to save certain animal species which have been identified as needing concerted protection effort. These projects are designed to protect the species *in situ*, by protecting and conserving their natural habitat – *Project Tiger* and *Project Elephant* are two such major initiatives. Other species conservation projects include *Project Crocodile Breeding and Management*, which was started in 1976 and now operates in 16 sanctuaries; and the Gir Lion Sanctuary Project in Gujarat, which aims at saving the Asiatic lion which once roamed over a wide stretch of the northern and central parts of the Indian subcontinent, but is today one of the most threatened species of animals in India. *Project Vulture, Project Snow-Leopard, Project Rhino, Project Hangul, Project Red Panda, Project Manipur Thamin, Project Olive Ridley Turtles* are also working in wildlife sanctuaries and national parks of India. According to *Forestry Statistics India 2011*, there are 668 Protected Areas (PAs) in India, of which 102 are national parks and 515 are wildlife sanctuaries. They include a variety of ecosystem and habitats. Some have been created in order to protect highly endangered species of wild plants and animals found nowhere else in world.

The Great Himalayan National Park is the largest sanctuary in this ecosystem and is one of the last homes of the beautiful snow leopard. The Dachigam Sanctuary is the only place where the rare Hangul of Kashmir stag is found. There are several sanctuaries in the *terai* region; the Kaziranga National Park is the most famous which has elephant, gaur, wild boar, swamp deer, and hog deer, in large numbers, as well as tigers and leopards. Its bird's life is extremely rich and includes ducks, geese, pelicans and storks. The Manas Sanctuary, in addition to the above *terai* species, also includes the rare golden langur and the very rare pygmy hog, the smallest wild boar in the world. The florican is found only in a few undisturbed grasslands in the *terai* sanctuaries.

In the Sal forest of Madhya Pradesh, there are several PAs. Kanha offers a wonderful opportunity to observe wild tigers. It is the only PA in which a subspecies of the *barasingha* (swamp deer) is found.

Bharatpur is one of the most famous water-bird sanctuaries in the world. Thousand of ducks, geese, herons, and other wading birds can be seen here. This is the only home of the very rare Siberian crane, which migrates to India every winter. During the last 20 years, the 30 or 40 Siberian cranes have dwindled to only 2 or 3.

In the Thar Desert, the wildlife is protected in the Desert National Park. Here, large numbers of blackbuck, *nilgai* and chinkara can be seen. The Great Indian Bustard lives in these arid lands.

Ranthambore is the most well-known sanctuary for observing tigers in the wild.

The Great and the Little Rann of Kutch have been made into sanctuaries to protect the very rare wild ass, the flamingo, the star tortoise and the desert fox.

In Gujarat, the Gir Sanctuary protects the last population of the majestic Asiatic lion. This thorn and deciduous forest is also the home of large herds of *chital, sambhar,* and *nilgai.*

The sanctuaries of the Western Ghats and associated hill ranges protect some of the most diverse forest types in the country. The few examples of highly threatened species include the Malabar

Box 2.4

Institutes Associated with the Protection of Biodiversity in India

1. The Indian Board of Wildlife: work under the Prime Minister. This is an apex body for guiding the implementation of various schemes of wildlife conservation.
2. The Animal Welfare Division is a part of Ministry of Environment and Forest.
3. National Institute of Animal Welfare (NIAW) at Faridabad.
4. Animal Welfare Board of India (AWBI) with Headquarters at Chennai.
5. The Committee for Purpose of Control and Supervision of Experiments on Animals (PCSEA) with Headquarters at Chennai.
6. Botanical Survey of India: established on 13 February 1890 with nine regional circles in the country.
7. Zoological Survey of India: established in 1916, has its Headquarters at Kolkata and has 16 regional stations.

giant squirrel, the flying squirrel and a variety of hill birds, several species of amphibians, reptiles and insects. These regions are also rich in highly endemic plant life. Sanctuaries such as Bhimashankar, Koyana, Chandoli and Radhanagari preserve this rich flora in Maharashtra; Bandipur, Bhadra, Dandeli, Nagarhole, etc. in Karnataka; and Eravikulam, Parambikulamm, Periyar, and the Silent Valley, in Kerala.

In the Nilgiri Hills the rich forest sanctuaries protect some of the last pockets of the Indian elephant in South India. Examples include Bandipur, Mudumalai, Wynad and Bhadra. During the last 10 years, a large number of the great tusker elephants of this region have been ruthlessly killed for their ivory. Now, very few of these magnificent animals are left in these jungles.

Two important sanctuaries meant for preservation of coastal ecosystem are the Chilka Lake in Odisha and Point Climere in Tamil Nadu. The Sunderbans protect the largest mangrove delta in India. The Marine National Park in Gujarat protects shallow areas in the sea, islands, coral reefs and extensive mudflats.

Over a hundred PAs have been created in the Andaman and Nicobar Islands to preserve their very special island ecosystem.

Integrated Protected Area System (IPAS): For the conservation of biodiversity, Protected Area is the effective system in every biogeographic region. A relatively larger representation must be included of highly fragile ecosystem areas for high species diversity or high endemism. PAs must also be integrated with each other, by establishing corridors between adjacent areas wherever possible, so that wildlife can move between them. The International Union for Conservation of Nature and Natural Resources (IUCN) states that it is essential to include at least 10 per cent of all ecosystems as Protected Areas if biodiversity is to be conserved in the long term.

India has only 5 per cent of lands in its 668 Protected Areas. However, much of this includes plantations of Sal or teak, which were developed for timber in the past and are thus relatively poor in diversity and have a low level of 'naturalness'. There are only a few good grasslands left in our country that are notified as PAs. Some are merely overgrazed wastelands in areas that were once flourishing grasslands. Most of these areas have a low biological value and need careful management to allow them to revert to a more 'natural' state, with their full complement of plants and animals.

Only a few wetlands have been made into sanctuaries. These require better management.

A major strategy to reduce impacts on the biodiversity of PAs should be to provide a sustainable source of resources for the local people living around them. A PA curtails their traditional grazing practices and access to fuelwood sources. These resources must be provided by developing them in the buffer areas. Plantations of fuelwood and good grassland management in areas outside PAs can help reduce the pressure on the habitat of wildlife within the PA. Good management practices must ensure that locals derive a direct economic benefit from the presence of the PA. Involving local people in Protected Area management and developing tourist facilities that support the income generation of the local people helps in enlisting their support for the PA.

(List of National Parks Sanctuaries, Tiger Reserves, Project Elephant Reserves and Biosphere Reserves in India is given at the end of this chapter).

Ex situ Conservation

Ex situ conservation means, conserving the species outside its natural habitat in a carefully controlled situation such as a botanical garden for plants or a zoological park for animals, where there is expertise to multiply the species under artificially-managed conditions. These breeding programmes for rare plants and animals are, however, more expensive than managing a PA.

There is also another form of preserving a plant by preserving its germplasm in a gene bank so that it can be used if needed in future. This is even more expensive. When an animal is on the brink of extinction, it must be carefully bred so that inbreeding does not lead to the genetic composition becoming weak. Breeding from the same stock can lead to poorly-adapted progeny or even an inability to be get enough offsprings.

Modern breeding programmes are done in zoos that provide for all the animal's needs, including enclosures that simulate their wild habitats. There may also be a need to assist breeding artificially.

A number of orchid sanctuaries, orchidariums, have been established in orchi-rich habitat, like the foothills of the Himalayas, the Western Ghats, the western coastal region and the southern hill stations.

The Central Council for Research in Ayurveda and Siddha (CCRAS), established in 1978, has promoted the cultivation of medicinal plants at five herbal gardens across the country covering an area of 135 acres.

The pygmy hog is the smallest and rarest wild pig in the world. Very small populations survive in the wild but it is almost on the brink of extinction. A small population exists in the Manas National Park in Assam. The Pygmy Hog Conservation Programme (PHCP), a collaborative programme of the government and international specialist groups, focuses its work on trying to ensure that pygmy hogs are successfully bred in captivity and released in the wild in an effort to save them from extinction.

In India, successful *ex situ* conservation programmes have been done for all our three species of crocodiles; this has been highly successful – example is the Madras Crocodile Trust Bank, which has grown from 10 to 8,035 crocodiles and the crocodiles here lay two clutches of eggs a year, compared to one in the wild. Another recent success has been the breeding of the very rare pygmy hog in the Guwahati Zoo. The Delhi Zoo has successfully bred the rare Manipur brow-antlered deer.

A lot of effort is underway to collect and preserve the genetic material of crop, animal, bird and fish species by the Government of India. This work is being done by institutions such as the National Bureau of Plant Genetic Resources (NBPGRs), New Delhi, and the National Bureau of Animal Genetic Resources (NBAGRs), Karnal.

Legislative Measures

Several legislative measures have been taken throughout the world to protect biodiversity. India has several Acts in force, which have a bearing on conservation of biodiversity. These are:

The Environment Protection Act, 1986, relates to general measures to protect the environment, such as restrictions on industrial and other processes or activities in specified areas. It also deals with the prevention of and control over the manufacture, use, release and movement of hazardous substances.

The Fisheries Act, 1897, prohibits the use of explosives and poison for fishing. It also regulates fishing in private waters.

The Forest Act, 1927, deals with the setting up and management of protected reserves and village forests, and controls on how selected products from the forest can be sold and at what prices.

The Forest (Conservation) Act, 1972; The Wildlife (Protection) Amendment Act, 1991; and The Wildlife (Protection) Amendment Act, 2002, deal with the restriction and prohibition of the hunting of animals, and with the protection of specified plants. They also deal with the setting up and management of sanctuaries and national parks, setting up of the Central Zoos Authority, control of zoos and captive breeding. They also control trade and commerce in wild animals, animal articles and trophies.

The Biodiversity Act, 2002, which was formulated after a long period of discussions and considerable public debate, is part of India's follow-up of the International Convention on Biological Diversity. The Act provides for a National Biodiversity Authority (NBA), which will screen proposals for the transfer of genetic resources abroad, and set up a system for genetic material from India to be sent to any other country. This is to ensure that there is a record of all genetic material going out of the country. The NBA will also advise the central government on measures for conservation, sustainable use, and benefit-sharing resulting from the use of the biodiversity with the communities who are its tradition custodians.

The National Biodiversity Strategy and Action Plan (NBSAP) was initiated in late 1999 by the Ministry of Environment and Forest (MoEF), Government of India, as part of India's commitment to the Convention on Biological Diversity (Biodiversity Convention). The broad purpose of this process was to produce an implementable action plan that would ensure the conservation of India's biodiversity, its sustainable use and equitable sharing of benefits arising from its use, i.e. the main tenets of the Biodiversity Convention. The two prerequisites of this plan are the ecological security of the country or of any region within it, and the livelihood security of those most critically dependent on biodiversity and its components.

Other efforts for the conservation of biodiversity in India are:

(i) Project coastal 'Mangroves for Future' (MFF) for coastal ecosystem conservation.
(ii) Four major coral reef areas identified for intensive conservation are Gulf of Mannor, Gulf of Kutch, Lakshadweep and Andaman and Nicobar Islands.

(iii) UNDP-GOICCF II Programme on promoting conservation of medicinal plants and traditional knowledge.
(iv) Wildlife Crime Control Bureau and National Board for Wildlife, etc.
(v) Biosphere Reserve Programme is an effective method of conservation of Abiotic and biodiversity of natural habitats. Biosphere Reserves are recognized within the framework of UNESCO's Man and Biosphere Programme. The world major ecosystem types network is devoted to conserving biological diversity. India has designated 17 Biosphere Reserves in different biogeographical regions of the country, out of which seven have been included in the world network of biosphere reserves of UNESCO.

Evolution of the Biodiversity Regime

The Stockholm Conference was the first occasion at which issues related to biodiversity were placed on the international agenda. A variety of legal instruments already existed. Some of these were regional, such as the Convention on Nature Protection Wildlife Preservation in the Western Hemisphere (Washington, D.C., 1940) and the African Resources (Algiers, 1968). Other measures dealt with specific species, such as birds and marine and polar region species. Prior to Stockholm, the only major international legal instrument for conserving biodiversity was the 1971 Convention on Wetlands of International Importance, especially for conservation of water flow habitats. The 61 parties to this convention pledged to make an inventory of designated wetlands, to promote their conservation, and establish nature reserves in the important wetlands.

But, no comprehensive international treaty addressed the problem of biological diversity. Although Stockholm did not produce such a document, it was for the first time that the issue was addressed in such a holistic manner. The issue of biodiversity was debated upon, and the conference documents included provisions aimed at protecting biodiversity.

The area of genetic resources and biotechnology was already controversial in 1972, at the time of the Stockholm Conference. Developing countries were concerned about their lack of access to resources, and they tried to use linkage politics to gain access to this technology. They pressed their demand for the transfer of technology on non-commercial terms. This demand was rejected by the developed countries, and the final statement of the conference diluted the language that addressed this linkage between economics and the environment.

Both the declaration of the United Nations Conference on the Human Environment and the Action Plan for the Human Environment included points which addressed biodiversity. Principles 2, 4 and 6 of the declaration included provisions for safeguarding flora and fauna and preserving ecosystems. The Action Plan generally addressed the importance of protecting forests, which among other things, were important habitats for wildlife. It recommended that wildlife be monitored to assess the impact of pollutants and that the economic value of wildlife be assessed. Governments were asked to take steps to protect ecosystems that have international significance. Recommendations 39 through 45 deal generally with the issue of preserving genetic resources.

In the period between the Stockholm and Rio conferences, the body of legislation concerning biodiversity continued to grow. Relevant major international legislations included the Convention concerning the Protection of the World Culture and Natural Heritage (1972), the Convention on International Trade in Endangered Species of Wild Fauna and Flora (1973), and Convention on Conservation of Migratory Species of Wild Animals (1979).

In the aftermath of Stockholm, developing countries continued their push for greater access to the biotechnology that resulted from their biological wealth. A non-binding International Undertaking on Plant Genetic Resources [Resolution (8/33)] was passed by the FAO in 1983. It stated that access to genetic resources should not be restricted, and it declared that all plant genetic resources were the common heritage of humanity; therefore, both primitive stocks and those developed by proprietary means should be free to all. Some developed world producers were opposed to this, and many industrialized countries were not parties to it because it was incompatible with their patent rights. Third World members, however accepted its provision.

The years between Stockholm and Rio saw incremental progress with regard to the items on the Third World's biodiversity agenda. Developing states wanted to make their own arrangements concerning resources access, and as the Merck deal illustrates, the slow process of norm change had begun. But, the progress of these states in the campaign for transfer of technology and financial assistance is more difficult to assess. Although developing countries maintained their determined resistance, these conflicting perspectives were reflected in the preparatory negotiations for the Rio conference.

The declaration of UN Conference on the Human Environment and the Action Plan for the Human Environment addressed the importance of protecting forests, which among other things, are important habitats for wildlife. Other conventions of biodiversity include May 1992 at Nairobi; June 1992 at UNCED; while in December 1993, Convention on Biological Diversity entered into force. The highlights of the conventions were:

(i) National identification and monitoring of biological diversity.
(ii) The development of national strategies and programmes for conserving biological diversity.
(iii) National *in situ* and *ex situ* conservation measures.
(iv) Environmental assessment procedures to take into account the effects of projects on biological diversity.
(v) National reports from parties on measures taken to implement in the convention and the effectiveness of the measures.

International Conservation Strategies

The issue of biodiversity is not limited to any country but it is a major international issue. Several international treaties and agreements are in place in the attempt to strengthen international participation and commitment towards conserving biodiversity. This is because certain problems can be addressed only if several countries work together. International agreement and treaties help in tackling such problems. Major international efforts done in this direction are as follows:

The Convention on Biological Diversity: This convention was signed during the United Nations Conference on Environment and Development (Earth Summit) in Rio de Janeiro, Brazil, in 1992. It focuses not only on conserving biodiversity but also on the sustainable use of biological resources and equitable sharing of benefits arising from its use. By signing this convention, member countries are committed to several activities including developing strategies for biodiversity conservation and sustainable use, and incorporating biodiversity issues into national plans, programmes and policies. This is by far the most important convention for the conservation of biodiversity at national and international levels.

The Convention on International Endangered Species of Wild Flora and Fauna (CITES): This convention is an international treaty which is designed to protect wild plants and animals affected by international trade. The treaty, in force since 1975, controls the export, import and re-export of endangered and threatened wildlife. Today, this convention provides varying degrees of protection to more than 30,000 species of animals and plants that are being traded as live specimens, for fur coats, or even as dried herbs.

The Convention on Wetlands of International Importance: This convention, also known as the Ramsar Convention, was signed in Ramsar (Iran) in 1971, and came into force in December 1975. It provides a framework for international cooperation for the conservation of wetland's habitats which have been designated to the List of Wetlands of International Importance. Although it originally focused on conserving habitats for waterbirds, the convention now covers all aspects of wetland conservation and their wise use. India became a signatory to this convention in 1982.

Table 2.3 *Ramsar Wetlands and their Sites in India*

Wetland	*State*
Kolleru Lake	Andhra Pradesh
Deepor Beel	Asom (Assam)
Nalsarovar	Gujarat
Pong Dam Lake	Himachal Pradesh
Renuka Wetlands	Himachal Pradesh
Chandertal Wetland	Himachal Pradesh
Hokera Wetland	Jammu and Kashmir
Surinsar-Mansar Lake	Jammu and Kashmir
Tsomoriri	Jammu and Kashmir
Wular Lake	Jammu and Kashmir
Thrissur Kole	Kerala
Vembanad-Kol Wetland	Kerala
Sasthamkotta Lake	Kerala
Ashtamudi Wetland	Kerala
Bhoj Wetland	Madhya Pradesh
Loktak Lake	Manipur
Chilka	Odisha
Bhitarkanika Mangrove	Odisha

Cont'd...

...Cont'd

East Calcutta Wetlands	West Bengal
Ropar	Punjab
Kanjili	Punjab
Harika Lake	Punjab
Keoladeo National Park	Rajasthan
Samber Lake	Rajasthan
Pointa Calimere Wildlife and Bird Sanctuary	Tamil Nadu
Rudrasagar Lake	Tripura
Upper Ganga River (Brijghat to Narora Stretch)	Uttar Pradesh

The World Heritage Convention: Also the Convention Concerning the Protection of the World Cultural and Natural Heritage, to give this international treaty its full name, is aimed at protecting sites of such outstanding value that their conservation is of concern to all people. Natural or cultural sites, which are nominated by the signatory nations, are evaluated for world heritage list. The treaty was adopted in Paris in 1972 and came into force in 1975. India has a total of 23 designated world heritage sites of which six are natural sites. These are Keoladeo National Park (Rajasthan), Manas National Park (Assam), Kaziranga National Park (Assam), The Sunderbans (West Bengal), Nanda Devi National Park (Uttar Pradesh), and Western Ghats of Maharashtra, Goa, Karnataka, Kerala and Tamil Nadu.

Worldwide Fund for Nature (WWF): It is an international non-governmental organization working on issues regarding conservation, research and restoration of environment. This organization also focuses on conservation of there biomes that contain most of world's biodiversity, i.e. forests, freshwater ecosystems, oceans, and coasts.

The Coalition Against Wildlife Trafficking (CAWT): It is a coalition of governments and international business and conservation organizations, working together to end the illegal trade in wildlife and wildlife products. India has also joined this organization.

The 'Tiger Summit': This took place in St. Petersburg in Russia in November 2010 with representatives of 13 Tiger Range States, the partner organization of the World Bank's Global Tiger Initiative. Since the Summit, the government has held two meetings in 2012, one at New Delhi in May and other in Bhutan in October.

Cartagena Protocol on Biosafety: On 29 January 2000, Biological Diversity Organization adopted an agreement known as the Cartagena Protocol on Biosafety. It entered into force on 11 September, 2003. The protocol was signed by 157 countries. The objective of this is to protect biological density from the potential risks posed by modified living organisms resulting from modern biotechnology.

International Treaty on Plant Genetic Resources for Food and Agriculture, 2004: This Treaty covers all plant genetic resources for food and agriculture, specifically, covers 64 crops. The main objectives of the treaty are conservation and sustainable use of plant genetic resources and equitable sharing of benefits arising out of their use.

Aichi Biodiversity Targets

In 2010, at the 10th Conference of the Convention on Biological Diversity (CBD) held at Nagoya (Japan), a new strategic plan for Biodiversity 2011–20 was adopted. The goals of 'Aichi' Biodiversity targets are:

1. Address the causes of biodiversity loss by mainstreaming biodiversity across government and society.
2. Reduce the direct pressure on biodiversity and promote sustainable use.
3. To improve the status of biodiversity by safeguarding ecosystem, species and genetic diversity.
4. Enhance the benefits of all, from biodiversity and ecosystem services.
5. Enhance implementation through participatory planning, knowledge management and capacity building.

National Parks and Sanctuaries in India

Name	*Location*	*Reserve for*
Namdapha Sanctuary	Arunachal Pradesh	Elephant, panther, sambhar, tiger, cheetal, king cobra.
Kaziranga National Park	Assam	One-horned rhinoceros, gaur, elephant, leopard, wild buffalo.
Sonai Rupa Sanctuary	Assam	Elephant, sambhar, wild boar, one-horned rhinoceros.

Cont'd...

...Cont'd

Gautam Buddha Sanctuary	Bihar	Tiger, leopard, sambhar, cheetal, barking deer.
Achanakmar Sanctuary	Chhattisgarh	Tiger, boar, cheetal, sambhar, bison.
Valvadore National Park	Gujarat	Wolf, black buck.
Wild Ass Sanctuary	Gujarat	Wild ass, wolf, nilgai, chinkara.
Gir Forest	Gujarat	India's biggest wildlife sanctuary famous for gir lions.
Nagarhole National Park	Karnataka	—
Dendeli Sanctuary	Karnataka	Tiger, panther, elephant, cheetal, sambhar, wild boar.
Bhadra Sanctuary	Karnataka	Elephant, cheetal, panther, sambhar, wild boar.
Banerghatta National Park	Karnataka	Elephant, cheetal, deer, grey patridge, green pigeon.
Bandipur Sanctuary	Karnataka and Tamil Nadu	Elephant, tiger, panther, sambhar, deer, birds.
Ranganthittoo Bird Sanctuary	Karnataka	Important bird sanctuary.
Tungabhadra Sanctuary	Karnataka	Panther, cheetal, sloth bear, four-horned antelope.
Dechigam Sanctuary	Kashmir	Kashmiri stag.
Pachmarhi Sanctuary	Madhya Pradesh	Tiger, panther, boar, sambhar, nilgai, barking deer.
Gandhi Sagar Sanctuary	Madhya Pradesh	Cheetal, sambhar, chinkara, wild birds.
Bandhavgarh National Park	Madhya Pradesh	Tiger, panther, cheetal, nilgai, wild boar.
Simlipal Sanctuary	Odisha	Elephant, tiger, leopard, gaur, cheetal.
Sunderban Tiger Reserve	West Bengal	Tiger, deer, wild boar, crocodile, Gangetic dolphin.
Jaldapara Sanctuary	West Bengal	Rhinoceros.
Ghana Bird Sanctuary	Rajasthan	Water birds, black duck, cheetal, sambhar.
Kangchenjunga National Park	Sikkim	Snow leopard, musk deer, Himalayan bear.
Vedanthangal Bird Sanctuary	Tamil Nadu	Important bird sanctuary.
Dudhwa National Park	Uttar Pradesh	Tiger, panther, sambhar, cheetal, nilgai, barking deer.
Chandraprabha Sanctuary	Uttar Pradesh	Gir lions, cheetal, sambhar.
Corbett National Park (named in memory of Jim Corbett)	Uttarakhand	Tiger, leopard, elephant, sambhar.

Tiger Reserves in India

S.No.	*State*	*Tiger Reserve*	*Remarks*
1.	Andhra Pradesh	Nagarjunsagar-Srisailam Tiger Reserve	Low tiger density.
2.	Andhra Pradesh	Kawal Tiger Reserve	20 tigers (2010).
3.	Arunachal Pradesh	Pakhui Tiger Reserve	Good tiger density.
4.	Arunachal Pradesh	Namdapha Tiger Reserve	Low tiger density, extreme encroachment by Lisus tribal people.
5.	Assam	Nameri Tiger Reserve	Encroachment and tree cutting by locals.
6.	Assam	Manas Tiger Reserve	Poor tiger density.
7.	Assam	Kaziranga Tiger Reserve	Good tiger density.
8.	Bihar	Valmiki Tiger Reserve	Low tiger density.
9.	Chhattisgarh	Indravati Tiger Reserve	—
10.	Chhattisgarh	Udanti & Sitanadi Tiger Reserve	Low tiger density.
11.	Chhattisgarh	Guru Ghasidas National Park	Separated from Sanjay National Park in Madhya Pradesh.
12.	Chhattisgarh	Achanakmar Tiger Reserve	Low tiger density.
13.	Goa	Mhadei Tiger Reserve	—
14.	Jharkhand	Palamau Tiger Reserve	Low tiger density.
15.	Karnataka	Bandipur Tiger Reserve	Good tiger density.
16.	Karnataka	Nagarhole (Extension) Tiger Reserve	Good tiger density.
17.	Karnataka	Dandeli-Anashi Tiger Reserve	Low tiger density.
18.	Karnataka	Bhadra Tiger Reserve	Low tiger density.
19.	Karnataka	Bannerghatta Tiger and Lion Reserve	Not a Project Tiger reserve.
20.	Karnataka	Biligiri Rangaswamy Temple Wildlife Sanctuary	17 tigers (2005).
21.	Karnataka	Kudermukh Tiger Reserve	20 tigers (2010).
22.	Kerala	Periyar Tiger Reserve	There are an estimated 53 tigers (2010) in the reserve.
23.	Kerala	Parambikulam Tiger Reserve	Moderate tiger density.
24.	Kerala, Tamil Nadu	Annamalai Tiger Reserve	Moderate tiger density.

Cont'd...

...Cont'd

25.	Madhya Pradesh	Bandhavgarh Tiger Reserve	Good tiger density.
26.	Madhya Pradesh	Bori-Satpura Tiger Reserve	Poor tiger density.
27.	Madhya Pradesh	Kanha Tiger Reserve	Good tiger density.
28.	Madhya Pradesh	Panna Tiger Reserve	—
29.	Madhya Pradesh	Pench Tiger Reserve	Good tiger density.
30.	Madhya Pradesh	Sanjay Dubri Tiger Reserve	Very low tiger density.
31.	Madhya Pradesh	Ratapani Tiger Reserve	—
32.	Maharashtra	Melghat Tiger Reserve	Low tiger density.
33.	Maharashtra	Pench Tiger Reserve	Moderate tiger density.
34.	Maharashtra	Tadoba-Andhari Tiger Reserve	Moderate tiger density, high rate of man-tiger conflicts.
35.	Maharashtra	Sahyadri Tiger Reserve	9 tigers (2007).
36.	Maharashtra	Nagzira-Navegaon Tiger Reserve	—
37.	Maharashtra	Bor Tiger Reserve	—
38.	Mizoram	Dampa Tiger Reserve	Low tiger density, due to ecological reasons.
39.	Odisha	Simlipal Tiger Reserve	Low tiger density.
40.	Odisha	Satkosia Tiger Reserve	Low tiger density.
41.	Odisha	Sunabeda Tiger Reserve	—
42.	West Bengal	Buxa Tiger Reserve	Low tiger density.
43.	West Bengal	Sunderbans Tiger Reserve	Good tiger density.
44.	Rajasthan	Ranthambore Tiger Reserve	Good tiger density.
45.	Rajasthan	Sariska Tiger Reserve	—
46.	Rajasthan	Mukundara Hills Tiger Reserve	—
47.	Tamil Nadu	Kalakad-Mundathurai Tiger Reserve	Low tiger density due to ecological reasons.
48.	Tamil Nadu	Mudumalai National Park	Moderate tiger density.
49.	Tamil Nadu	Sathyamangalam Tiger Reserve (Pro)	46 tigers (2010).
50.	Uttar Pradesh	Dudhwa Tiger Reserve	Good tiger density.
51.	Uttar Pradesh	Pilibhit Tiger Reserve	—
52.	Uttar Pradesh	Suhelwa Tiger Reserve	—
53.	Uttar Pradesh	Corbett Tiger Reserve	Good tiger density.

Elephant Reserves in India

Elephant Range	Elephant Reserve	State
Eastern India (South-West)	1. Mayurjharna	West Bengal
Bengal-Jharkhand-Odisha	2. Singhbhum	Jharkhand
	3. Mayurbhanj	Odisha
	4. Mahanadi	Odisha
	5. Sambalpur	Odisha
	6. Baitarni	Odisha
	7. South Orrisa	Odisha
	8. Lemru	Chhattisgarh
	9. Badalkhol Tamorpingla	Chhattisgarh
North Brahamputra	10. Kamerng	Arunachal
Arunachal Pradesh	11. Sonitpur	Assam
South Brahamputra	12. Dihing-Patkai	Assam
Assam, Arunachal	13. Deomoli	Arunachal
Kaziranga	14. Kazinranga-Karbi Anglong	Assam
Assam	15. Dhansiri-Lungding	Assam
Nagaland	16. Intanki	Nagaland
Eastern Doars	17. Chirang Ripu	Assam
Assam-West Bengal	18. Eastern Dooars	West Bengal
East Himalaya	19. Garo Hills	Meghalaya
Meghalaya	20. Khasi Hills	Meghalaya
Nilgiri-Eastern Ghats	21. Mysore	Karnataka
Karnataka-Kerala	22. Wayanad	Kerala
Tamil Nadu	23. Nilgiri	Tamil Nadu
Andhra Pradesh	24. Rayala	Andhra Pradesh
South Nilgiri	25. Nilambur	Kerala
Kerala-Tamil Nadu	26. Coimbatore	Tamil Nadu
Western Ghats	27. Annamalai	Tamil Nadu
Tamil Nadu-Kerala	28. Anaimudi	Kerala
Periyar	29. Periyar	Kerala
Kerala-Tamil Nadu	30. Srivilliputtar	Tamil Nadu
Northern India	31. Shiwalik	Uttarakhand
Uttarakhand	32. Uttar Pradesh	Uttar Pradesh

Biosphere Reserves in India

Reserves	Location	State	Type	Key Fauna
Seshachalam Hills	Seshachalam hill ranges covering parts of Chittoor and Kadapa districts	Andhra Pradesh	Eastern Ghats	—
Great Nicobar Biosphere Reserve	Southern most Islands of Andaman and Nicobar	Andaman and Nicobar Islands	Islands	Saltwater Crocodile
Dihang-Dibang	Part of Siang and Dibang Valley	Arunachal Pradesh	Eastern Himalayas	—
Manas	Part of Kokrajhar, Bongaigaon, Barpeta, Nalbari, Kamrup and Darrang district	Assam	Eastern Himalayas	Golden Langur, Red Panda
Dibru-Saikhonwa	Part of Dibrugarh and Tinsukia districts	Assam	Eastern Himalayas	Golden Langur
Rann of Kutch	Part of Kutch, Rajkot, Surendranagar and Patan districts	Gujarat	Desert	Indian Wild Ass
Cold Desert	Pin Valley National Park and surrounding Chandratal and Sarchu and Kibber Wildlife Sanctuary	Himachal Pradesh	Western Himalayas	Snow Leopard
Agasthyamalai Biosphere Reserve	Neyyar, Peppara and Shenduruny Wildlife Sanctuary and their adjoining areas	Kerala, Tamil Nadu	Western Ghats	Nigiri Tahr, Elephants
Panna	Part of Panna and Chattarpur districts	Madhya Pradesh	Catchment area of Ken River	Tiger, Chital, Chinkara, Sambhar, Sloth bear
Pachmarhi Biosphere Reserve	Parts of Betul district, Hoshangabad district and Chhindwara district	Madhya Pradesh	Semi-Arid	Giant Squirrel, Flying Squirrel
Achanakamar Amarkantak	Part of Annupur, Dindori and Bilaspur districts	Madhya Pradesh, Chhattisgarh	Maikala Range	—
Nokrek	Part of Garo Hills	Meghalaya	East Himalayas	Red Panda
Simlipal	Part of Mayurbhanj districts	Odisha	Deccan Peninsula	Gaur, Royal Bengal Tiger, Wild elephant

Cont'd...

...Cont'd

Sunderbans	Part of delta of Ganges and Brahmaputra river system	West Bengal	Gangetic Delta	Royal Bengal Tiger
Kangchendzonga	Parts of Kangchanjunga Hills	Sikkim	East Himalayas	Snow Leopard, Red Panda
Nilgiri Biosphere Reserve	Part of Waynad, Nagarhole, Bandipur and Mudumalai, Nilambur, Silent Valley and Siruvani Hills	Tamil Nadu, Kerala and Karnataka	Western Ghats	Nilgiri Tahr, Lion-tailed macaque
Gulf of Mannar	Indian part of Gulf of Mannar extending from Rameshwaram in the North to Kanyakumari in the South of Tamil Nadu and Sri Lanka	Tamil Nadu	Coasts	Dugong or Sea Cow
Nanda Devi National Park & Biosphere Reserve	Part of Chamoli, Pithoragarh and Bageshwar districts	Uttarakhand	Western Himalayas	—

MULTIPLE-CHOICE QUESTIONS

1. Biodiversity refers to:

   (a) Diversity of soil

   (b) Diversity of landscape

   (c) Diversity of the species of living organisms

   (d) Socio-cultural and economic diversities

2. Who first used the term 'biological diversity'?

   (a) W.G. Rosen (b) Ramond F. Dasmann

   (c) Thomas Loverjoy (d) E.O. Wilson

3. Which is not a type of biodiversity?

   (a) Hydro diversity (b) Genetic diversity

   (c) Species diversity (d) Ecosystem diversity

4. Which are the categories used for the mapping of biodiversities in the world?

   (a) Alpha diversity

   (b) Beta Diversity

   (c) Gama diversity

   (d) All these

5. The greatest threat to the world's coral reef systems emanates from:

   (a) UV light band
   (b) Starvation caused by decline in zooplankton
   (c) Cyanide fishing
   (d) More frequent coral bleaching

6. The greatest threat to biodiversity?

   (a) Process of habitat loss
   (b) Species extinction all over the world
   (c) Reduced carrying capacity of the habitat
   (d) Biodiversity hotspots

7. In a mountainous terrain rich biodiversity is found in areas located at:

   (a) Lower latitudes and higher altitudes
   (b) Higher latitudes and lower altitudes
   (c) Lower latitudes and lower altitudes
   (d) Higher latitudes and higher altitudes

8. **Assertion (A):** Eutrophication disrupts natural functioning of the ecosystem bringing about negative environment effects.

   Reason (R): Enhanced growth of phytoplankton and algal blooms leads to lack of oxygen in the water body resulting in severe reduction in fish and other animal populations.

   Examine the above statements and select correct answer from the codes given below:

   (a) Both **A** and **R** are true and **R** is the correct explanation of **A**
   (b) Both **A** and **R** are true and **R** is not the correct explanation of **A**
   (c) **A** is true but **R** is false
   (d) **A** is false but **R** is true

9. Consider the following statements:

   1. Biodiversity is normally greater in the lower latitudes as compared to the higher altitudes.
   2. Along the mountain gradients biodiversity is normally greater in lower latitudes as compared to the higher latitudes.

   Which of the statements given above is/are correct?

   (a) only 1 (b) only 2

   (c) both 1 and 2 (d) neither 1 nor 2

10. Biodiversity forms the basis for human existence in the following ways:

   1. Soil formation 2. Preservation of soil erosion

   3. Recycling of waste 4. Pollination of crops

   Select the correct answer using the codes given below:

   (a) 1, 2 and 3 (b) 2, 3 and 4

   (c) 1 and 4 (d) all of these

11. Which one of the following is not a site for *in situ* method conservation of flora?

   (a) Biosphere Reserve

   (b) Botanical Garden

   (c) National Park

   (d) Wildlife Sanctuary

12. The Himalayan ranges is very rich in species diversity. Which one among the following is the most appropriate reason for this phenomena?

   (a) It has a high rainfall that supports luxuriant vegetation

   (b) It is a confluence of different biogeographical zones

   (c) Exotic and invasive species have not been introduced in this region

   (d) It has less human interference

13. Match the following lists:

	List I (Biosphere Reserve)		List II (State)
A.	Simplipal	1.	Sikkim
B.	Dihang Dibang	2.	Uttarakhand
C.	Nokrek	3.	Arunachal Pradesh
D.	Kangchendzonga	4.	Odisha
		5.	Meghalaya

Codes:

	A	**B**	**C**	**D**
(a)	1	3	4	5
(b)	4	5	2	1
(c)	1	5	2	4
(d)	4	3	5	1

14. Which one among the following has the maximum number of national parks?

(a) Andaman and Nicobar Islands
(b) Arunachal Pradesh
(c) Assam
(d) Meghalaya

15. Which one of the following is not a biosphere reserve?

(a) Agasthyamalai (b) Nallamalai
(c) Nilgiri (d) Pachmarhi

16. Which one of the following is located in Bastar region?

(a) Bandhavgarh National Park
(b) Randeli Sanctuary
(c) Rajaji National Park
(d) Indrawati National Park

17. Consider the following statements:

1. Silent Valley National Park is in the Nallamalai
2. Pathrakavedu hydroelectric plant project is proposed to be build near the Silent Valley National Park
3. The Kunthi river originates in Silent Valley rainforest.

Which of the statements given above is/are correct?

(a) 1 and 2 (b) only 2
(c) 2 and 3 (d) All of these

18. Consider the following regions:

1. Eastern Himalayas
2. Eastern Mediterranean region
3. North-West Australia

Which of the above is/are biodiversity hotspot(s)?

(a) only 1 (b) 1 and 2
(c) 2 and 3 (d) All of these

19. Consider the following areas:

1. Bandipur 2. Bhitarkanika
3. Manas 4. Sunderbans

Which of the above are tiger reserves?

(a) Both 1 and 2 (b) 1, 3 and 4
(c) 2, 3 and 4 (d) All of These

20. Which of the following is not a Ramsar Site?

(a) Bhitankanika mangroves (b) Chilika lake
(c) Deepor Bul (d) All the above

21. Which one of the following is not a biogeographic zone of India?

(a) Chhota-Nagpur Plateau
(b) Thar Desert

(c) Western Ghats
(d) Gangetic Plain

22. Which among the following is not a hotspot located in tropical zone?

(a) Ivory Coast (b) Central Chile
(c) Tanzania (d) Northern Borneo

23. Which among the following is not listed in 'Red Data Book'?

(a) Endangered species (b) Extinct species
(c) Flourishing species (d) Vulnerable species

24. In which year International Union for Conservation of Nature and Natural Resources was established?

(a) 1980 (b) 1988
(c) 1985 (d) 1990

25. Which among the following is not an elephant reserve?

(a) Mayurbhanj (b) Deomoli
(c) Panna (d) Rayales

26. Match the following lists:

List I (Wildlife Sanctuaries)	List II (State)
A. Namdapha	1. Karnataka
B. Periyar	2. Arunachal Pradesh
C. Bandipur	3. Manipur
D. Lamjao	4. Kerala

Codes:

	A	**B**	**C**	**D**
(a)	4	2	3	1
(b)	2	4	1	3
(c)	4	2	1	3
(d)	2	4	3	1

27. Seacoast of which one of the following states has become famous as a nesting place for the giant Olive Ridley sea turtles from South America?

(a) Goa (b) Gujarat

(c) Odisha (d) Tamil Nadu

28. Consider the following organisms:

1. Bacteria
2. Fungi
3. Flowering plants

Some species of which of the above kinds of organisms are employed as bio-pesticides?

(a) Only 1 (b) 2 and 3

(c) 1 and 3 (d) All of these

29. Consider the following statements:

1. Biodiversity hotspots are located only in tropical regions.
2. India has four biodiversity hotspots, i.e. Eastern Himalayas, Western Himalayas, Western Ghats and Andaman and Nicobar Islands?

Which of the statements given above is/are correct?

(a) Only 1 (b) Only 2

(c) Both 1 and 2 (d) Neither 1 nor 2

30. The International Convention on Wetlands was held at Ramsar. In which country is Ramsar located?

(a) Turkey (b) Sri Lanka

(c) India (d) Iran

31. Ramsar International Convention is related to:

(a) Forest Conservation (b) Wetlands

(c) Soil Conservation (d) All of these

32. What is true about the Asan Wetlands in Uttarakhand?

(a) It has been declared a Biosphere Reserve

(b) It is one the conservation reserves

(c) It is known as the original homeland of river\dolphins

(d) It is a famous nesting ground for hill myanas.

33. Consider the following pairs:

1. Bhiterkanika Orissa – Salt Water Crocodile
2. Desert National Park, Rajasthan – Great Indian Bustard
3. Eravikulam, Kerala – Hoolak Gibbon

Which of these pairs given above is/are correctly matched?

(a) only 1 (b) 1 and 2 only

(c) only 2 (d) 1, 2 and 3

34. Which one of the following groups of animals belongs to the category of endangered species?

(a) Great Indian Bustard, Musk deer, Red Panda and Asiatic Wild Ass.

(b) Kashmir stag, Cheetal, Blue Bull and Great Indian Bustard.

(c) Snow Leopard, Swamp Deer, Rhesus Monkey and Saras (Crane).

(d) Lion tailed Macaque, Blue Bull, Hanuman Langur and Cheetal.

35. Consider the following pairs:

1. Indravati Tiger Reserve – Chhattisgarh
2. Kundermukh Tiger Reserve – Karnataka
3. Ratapani Tiger Reserve – Maharashtra
4. Satkosia Tiger Reserve – Odisha

Which of these pair given above is/are correctly matched?

(a) Only 1 (b) Only 2

(c) 2 and 3 (d) 1, 2 and 4

36. Which of the following Tiger Reserve is located in Rajasthan?

(a) Suhelwar (b) Sariska
(c) Bandipur (d) Valmiki

37. Consider the following statements:

1. The boundaries of National Park are defined by legislation
2. A biosphere reserve is declared to conserve a few species of flora and fauna
3. In a wildlife century, limited biotic interference is permitted

Which of the statements given above is/are correct?

(a) Only 1 (b) 2 and 3
(c) 1 and 3 (d) All of these

38. Which one of the following Union ministries implement the Contagena Protocol on Biosafety?

(a) Ministry of Science and Technology
(b) Ministry of Health and Family Welfare
(c) Ministry of Environment and Forest
(d) Ministry of Chemical and Fertilizers

39. Match the following lists

	List I (Mangrove)	List II (State)
A.	Achra Ratnagir	1. Karnataka
B.	Coondapur	2. Kerala
C.	Pichararam	3. Andhra Pradesh
D.	Vembanad	4. Maharashtra
		5. Tamil Nadu

Codes:

	A	B	C	D
(a)	2	1	5	4
(b)	4	5	3	2
(c)	2	5	3	4
(d)	4	1	5	2

40. Out of all the Biosphere Reserves in India, four have been recognized on the world network by UNESCO. Which one of the following is not one of them?

(a) Gulf of Mannar (b) Kangchanjunga
(c) Nanda Devi (d) Sunderbans

41. Which one of the following statements is/are correct?

(a) The first Meeting of the Parties (MoP) to Cartagerna protocol on biosafety was held in Philippines in the year 2004.
(b) India is not a signatory to the Biosafety Protocol convention on biological diversity.
(c) The Biosafety Protocol deals with genetically modified organisms.
(d) The United States of America is member of the Biosafety Protocol Convention of Biological Convention of Biological diversity.

42. Which of the following can be threats to the biodiversity of geographical area?

1. Global warming
2. Fragmentation of habitat
3. Invasion of alien species
4. Promotion of vegetarianism

Select the correct answer using the codes given below:

(a) 1, 2 and 3 (b) 2 and 3
(c) 1 and 4 (d) All of these

43. How does National Biodiversity Authority (NBA) help in protecting the Indian agriculture?

1. NBA checks the biopiracy and protects the indigenous and traditional genetic resources

2. NBA directly monitors and supervises the scientific research on genetic modification of crop plants
3. Application for Intellectual Property Rights related to genetical biological resources cannot be made without the approval of NBA

Which of the statements given above is/are correct?

(a) Only 1 (b) 2 and 3

(c) 1 and 3 (d) All of these

44. Consider the following statements:

1. Genetic diversity is essential for healthy breeding population of a species
2. If the number of breeding is reduced, this leads to genetic anomalies and eventually, to extinction of that particular species
3. The number of species of plants and animals in a region constitutes its biodiversity

Which of the above statements(s) is/are correct about genetic diversity?

(a) Only 1 (b) 1 and 2

(c) 1 and 3 (d) 2 and 3

45. Which is the largest coral reef ecosystem in the world?

(a) Coral reefs of Philippines

(b) Coral reefs of Indonesia

(c) Australian Great Barrier Reef

(d) Coral reef of Seychelles

46. Which among the following is not a wetlands?

(a) Marshes (b) Swamps

(c) Mangroves (d) Coastal Plains

47. Consider the following statements:

1. Biological diversity is essential for preserving ecological processes.
2. Biodiversity is the result of climatic change.
3. Biodiversity provides a variety of environmental services that are essential at global, regional and local levels.

Which of the above statement(s) is/are true for importance of biodiversity?

(a) Only 1 (b) Only 2
(c) 1 and 3 (d) All of these

49. The Convention on Biological Diversity was signed during

(a) UN Conference on Environment and Development at Rio de Janeiro
(b) At Stockholm Environmental Conference
(c) World Heritage Convention
(d) Cartagena Protocol on Biosafety

50. In which year Biodiversity Act was introduced in India?

(a) 1986 (b) 1990
(c) 1992 (d) 2002

51. Which of the following statements is/are part of approach under Project Tiger?

1. Elimination of all forms of human interference from the areas of tigers reserve
2. Protective management of habitat
3. Researching details about existing habitat and wild animals and carefully monitoring changes in flora and fauna

Select the correct answer using the codes given below:

(a) 1 and 2 (b) Only 3
(c) 2 and 3 (d) All of these

52. Consider the following statements about Convention on International Trade in Endangered Species (CITES):

1. It is also known as the Washington Convention
2. It is a legally binding agreement on the member nations/parties
3. Its aim is to ensure that international trade in specimens of wild animals and plants does not threaten the survival of the species in the wild

Select the correct statement(s) using the code given below:

(a) 1 and 2 (b) 1 and 3
(c) Only 3 (d) All of these

53. Consider the following statements regarding National Green Tribunal (NGT):

1. NGT was established on 18 October 2010.
2. Its main objectives are for effective and expeditious disposal of cases relating to environmental protection and conservation of forests.
3. It is a specialized body equipped with the necessary expertise to handle the environmental disputes.

Which of the statements given above is/are correct?

(a) 1 and 2 (b) 1 and 3
(c) 2 and 3 (d) All of these

54. Kaziranga National Park belongs to which one of the following biogeographical regions?

(a) Tropical Humid Forest (b) Warm Semi-Desert
(c) Tropical Dry Forest (d) Tropical Sub-humid Forest

55. Consider the following statements:

1. National Green Tribunal is a statutory body established under Legislative Act in 2010

2. Number of judicial members in National Green Tribunal ranges from 10–20.
3. Number of expert members in National Green Tribunal ranges from 10–20.

Select the correct answer using the codes given below:

(a) 1 and 2 (b) Only 3
(c) 2 and 3 (d) All of these

56. Biodiversity Act of 2002, India enshrined which one of the following?

(a) Conservation of biological resources
(b) Accessibility and equitable distribution of these resources
(c) Sustainability
(d) All of the above

57. Consider the following statements:

1. Nagarjunsagar-Srisailam is the largest tiger reserve in terms of area
2. Panda Tiger Reserve in Maharashtra is minimum in area
3. As per website of NTCA, Sunderbans in West Bengal has the largest number of tigers

Which of the above statements is/are incorrect?

(a) Only 1 (b) 1 and 2
(c) All of these (d) None of these

58. Gambhri river is related to which of the following National Park:

(a) Keoladeo (b) Corbett
(c) Bhiarkanika (d) Dachigam

59. In which biosphere reserve is Indian Wild Ass found?

(a) Simlipal (b) Rann of Kutch
(c) Manas (d) Panna

60. In which biosphere reserve is Red Panda found?

(a) Dehang-Dibang (b) Nilgiri
(c) Nokreak (d) Nanda Devi

61. Consider the following statements:

1. Aichi Biodiversity Targets was adopted in 10th Convention of Biological Diversity held at Nagoya
2. A new strategic plan for biodiversity was adopted for 2011–2020
3. Aichi Biodiversity Targets aims to protect the forests
4. Aichi Biodiversity Target aims to reduce the direct pressures on biodiversity and promote sustainable use

Which of the above statement(s) is/are correct for Aichi Biodiversity Targets?

(a) Only 1 (b) 1, 2 and 3
(c) 1, 2 and 4 (d) All of these

62. What was the objective of the Tiger Summit held in November, 2010?

(a) To have a world census of tigers
(b) To finalize a Global Tiger Recovery Programme
(c) To delimit the tiger territory in various parts of the world
(d) To develop possibility of tiger exchange between countries

63. In which year was the Forest (Conservation) Act passed in India?

(a) 1927 (b) 1950
(c) 1980 (d) 1986

64. The Wildlife (Protection) Act was passed in India in 1972. When amendment (s) in this act was/were done?

(a) 1986 (b) 1990 and 1992
(c) 1994 and 2000 (d) 1991 and 2002

65. Which plan was formulated in India, as a part of commitment to the Convention on Biological Diversity?

(a) Biosphere Reserve Programme
(b) National Biodiversity Strategy and Action Plan
(c) Monitoring of Biological Diversity Plan
(d) All of these

66. Botanical garden for plants and zoological park for animals are the examples of which type of biodiversity conservation?

(a) *Ex situ* (b) *In situ*
(c) Integrated System (d) Legislative systems

67. Which among the following has been designed to protect the species *in situ*?

(a) Wetland Project
(b) Tropical Forest Project
(c) Project Rhino
(d) All of these

68. Which among the following was the first National Park established in India?

(a) Bandhavgarh National Park
(b) Corbett National Park
(c) Dudhwa National Park
(d) Kaziranga National Park

69. Which sanctuary is famous for 'Kashmiri Stag'?

(a) Simlipal Sanctuary
(b) Dachigam Sanctuary
(c) Chandraprabha Sanctuary
(d) Bandipur Sanctuary

70. Which among the following is a correct pair?

(a) Gautam Buddha Sanctuary – Bihar
(b) Jaidapara Sanctuary – West Bengal
(c) Achanakmar Sanctuary – Karnataka
(d) Chandraprabha Sanctuary – Uttar Pradesh

71. Match List I with List II and select answer as per code given below the lists:

List I (Drugs)		List II (Plant Source)	
A.	Menthol	1.	Yellow cinchona
B.	Papain	2.	Cocoa
C.	Quinine	3.	Mint
D.	Cocaine	4.	Papaya

Codes:

	A	**B**	**C**	**D**
(a)	1	3	4	2
(b)	2	4	3	1
(c)	3	4	1	2
(d)	4	1	2	3

72. Which plant is the source of anticancer drug Toxol?

(a) Belludonna (b) Pacifice yew
(c) Thorn apple (d) Autumn Crocus

73. Which one of the following is not a 'hot spot' of the world?

(a) Tropical Andes (b) Gangetic
(c) South-Central China (d) Madagascar

74. Which of the following is not a main cause of the loss of biodiversity?

(a) Introduction of foreign species

(b) Habitat loss and fragmentation

(c) Export policy

(d) Climate change and pollution

Note: Question number 75 to 79 are based on Assertion (A) and Reason (R). Select your answer using the following codes:

(a) Both **A** and **R** are true and **R** is the correct explanation of **A**

(b) Both **A** and **R** are true but **R** is not the correct explanation of **A**

(c) **A** is true **R** is false.

(d) **A** is false **R** is true.

75. **Assertion (A):** Biodiversity is extremely complex, dynamic and varied like other features of the earth.

Reason (R): By changing biodiversity, we strongly affect human well-being and the well-being of every other living creature.

76. **Assertion (A):** Each member of any animal or plant species differs widely from other individuals in its genetic make up.

Reason (R): The genetic variability is essential for healthy breeding population of species.

77. **Assertion (A):** Biological diversity is essential for preserving ecological processes, such as of nutrients, soil formation, circulation and cleansing of air and water.

Reason (R): Biodiversity provides a variety of environmental services from its species and ecosystems that are essential at the global, regional and local levels.

78. **Assertion (A):** Tropical regions of the world are having highest biodiversity.

Reason (A): Tropical regions are having wet and warm climate.

79. **Assertion (A):** The extinction of several species or loss of biodiversity is a major problem of the international world.

Reason (R): Destruction of habitats, introduction of foreign species and hunting are responsible for the loss of biodiversity.

80. Match the following lists:

List I	List II
A. Gulf of Mannar	1. Assam
B. Pachmarhi	2. Sikkim
C. Dibru-Saikhonwa	3. Madhya Pradesh
D. Kanchenjunga	4. Tamil Nadu

Codes:

	A	**B**	**C**	**D**
(a)	1	3	4	2
(b)	4	3	1	2
(c)	3	2	4	1
(d)	2	1	3	4

81. Which of the following pair is not correctly matched?

Elephant Reserve		State
(a) Lemru	–	Chhattisgarh
(b) Deomoli	–	Arunachal Pradesh
(c) Nilambur	–	Kerala
(d) Anaimudi	–	Tamil Nadu

82. Match the following lists:

List I (Tiger Reserve)	List II (State)
A. Namndapha	1. Goa
B. Mhadei	2. Rajasthan
C. Kudermukh	3. Arunachal Pradesh
D. Ranthambore	4. Karnataka

Codes:

	A	B	C	D
(a)	2	3	1	4
(b)	1	2	4	3
(c)	3	1	4	2
(d)	4	2	3	1

83. Which of the following tiger reserve is not located in Madhya Pradesh?

(a) Suhelwa (b) Bandhavgarh
(c) Bori-Satpura (d) Ratapani

84. Which of the following is not a Bird Sanctuary in India?

(a) Ranganthitto (b) Pachmarhi
(c) Vedanthangal (d) Ghana

85. Consider the following statements:

1. Most of the world's bio-rich nations are in the South
2. Indian Ocean is having richest and most diverse coral reefs
3. Majority of the countries capable of exploiting biodiversity are northern nations
4. all the species are the integral part of ecosystem

Which of the above statement(s) is/are correct?

(a) 1 and 2 (b) 2 and 3
(c) 2, 3 and 4 (d) 1, 3 and 4

86. The development of different types of plants and animal varieties are included in which type of biodiversity?

(a) Microorganism Biodiversity
(b) Species diversity
(c) Domesticated Biodiversity
(d) Genetic Diversity

87. Which of the following statements is not correct regarding biodiversity?

(a) Biodiversity in countries like India meets the basic survival needs of a vast number of people

(b) Up to 80 per cent of the people in developing countries depend for primary health care on traditional medicines

(c) Biodiversity helps in reducing occurrence of earthquake

(d) Biodiversity is involved in water purification, recycling nutrients and providing fertile soil

88. Which of the following is not a Ramsar wetland site?

(a) Kolleru Lake (b) Wular Lake

(c) Rudrasagar Lake (d) Pushkar Lake

89. Which of the sanctuaries in India is famous for Rhinoceros?

(a) Sunderban (b) Jaldapara

(c) Pachmarhi (d) Dandeli

90. Which one of the following state in India is not having tiger reserves?

(a) Himachal Pradesh (b) Rajasthan

(c) Mizoram (d) Goa

91. Which are of the following is not a world heritage natural site in India?

(a) Kaziranga (b) Nanda Devi

(c) Kashmir Valley (d) Western Ghats

Answers

1. c	2. b	3. a	4. d	5. d	6. a	7. c
8. a	9. c	10. d	11. b	12. b	13. d	14. c
15. b	16. d	17. c	18. b	19. b	20. d	21. a
22. b	23. c	24. b	25. c	26. b	27. c	28. d
29. d	30. d	31. b	32. b	33. b	34. a	35. d
36. b	37. c	38. c	39. d	40. b	41. c	42. a
43. c	44. b	45. c	46. d	47. c	48. c	49. a
50. d	51. b	52. c	53. d	54. a	55. d	56. d
57. d	58. a	59. b	60. c	61. c	62. b	63. c
64. d	65. b	66. a	67. c	68. b	69. b	70. a
71. c	72. b	73. b	74. c	75. b	76. b	77. a
78. a	79. a	80. b	81. d	82. c	83. a	84. b
85. d	86. c	87. c	88. d	89. b	90. a	91. c

3 Climate Change

Climate change is the single biggest environmental and humanitarian crisis of our time. The earth's atmosphere is overloaded with heat-trapping carbon dioxide, which causes large-scale disruption in climate with disastrous consequences. Global climate change has been a burning problem for environmentalists, since the middle of the 20th century and has stirred up a serious debate among the researchers even in the contemporary scientific community. Small and Witherick (1995) have defined climate change as 'A change in the climate of a region or place, usually reflected in the mean annual temperatures or mean annual precipitation, over a time-period that can vary from long term to short term'. Thus, climate change refers to a large scale change in one or more basic climate components, such as temperature or precipitation. Such a change would be significant enough to have profound impact on the natural environment and human society.

Climate change refers to the variation in the earth's global climate or in regional climates over time. It describes change in the variability or average state of the atmosphere over a time scale ranging from decades to millions of years. These changes can be used by processes internal to the earth and external forces for example, variation in sunlight intensity or more recently, human activities. In recent usages, especially in the context of environment policy, the term 'climate change' often refers to changes in modern climate which according to the Intergovernmental Panel on Climate Change (IPCC) are 90 to 95 per cent likely to have been, in part, caused by human action. Consequently, the term anthropogenic climate change is frequently adopted; this phenomenon is also referred to commonly as 'global warming'. Thus, now the term climate change has become synonymous with 'global warming'. Global warming is the increase of earth's average gases, such as carbon dioxide emissions from burning fossils or deforestation, which trap heat that would otherwise escape from the earth. This is a type of greenhouse effect. This chapter presents the climate change in terms of global warming and greenhouse effects as main source of global warming/climate change. Both these aspects are interrelated with each other and are responsible for climate change.

Climate Change in the Past

The earth's temperature has not been constant in the past. Indeed our planet has gone through several periods of global cooling and global warming as it has moved in and out of what we call 'ice ages'. Variability, in both time and space, is an inherent feature of climate, as the atmosphere is always in the state of turmoil and instability leading to variations in weather and climatic conditions. The climochronology (history of palaeoclimates) reveals the fact that climate has changed in geological past and hence it is opined that 'the world's climates have changed in the past, are changing now, and there is every reason to expect that they will change in future' (J.E. Hobbs, 1980). The change in climates is a reality because climate of a region is not static rather it keeps changing, sometimes in a cyclic manner, known as climatic cycle. This has happened in the past, evident from the occurrences of ice ages during (i) Pre-Cambrian period (250–600 million year before present, mbp), (ii) Ordovician period (450–430 mbp), (iii) Carboniferous-Permian periods (300 mbp), (iv) Pleistocene period (2–3 mbp).

The final retreat and withdrawal of recent ice sheets started about 18,000 years before present and continued up to 10–12,000 years before present, when most of the ice covers were withdrawn from the USA. The record shows that the climatic conditions of Europe in 1st century of Christian era was similar to that of present conditions. Europe and North America experienced dry phase resulting into semi-drought conditions in the 5th century A.D. The climates in northern hemisphere became further harsh during 600–700 A.D.

The period from 95 to 1250 A.D. is called as 'Phase of Little Climatic Optinum' in the climatic history of earth, when climate became warm and relatively dry as average temperature increased by 1° to 2° C from the present day global average temperatures. The period from 1250 A.D. to 1450 A.D. was characterized by the reversal of mild climate of 10th to 13th centuries A.D. The climate continued to become further harsh for another 400 years (from 1450 A.D. to 1880 A.D.) as the temperature fell below freezing point, giving birth to another glacial period known as Little Ice Age. The coldest period culminated in the year 1876 when major parts of northern Europe and the USA did not experience summer season. The year 1816 is known as the 'the year without summer' in the climatic history. Climate began to improve due to rise in temperature and by the end of 19th century the ice age had terminated.

The records of global temperatures since Industrial Revolution (1860) denote slow but irregular trend of rise of temperature. R.G. Barry and R.J. Chorlay (2002) have identified the following trends in rise of temperatures after 1860:

- Highly irregular trend of rise in temperature ranging between 0.2°C to 0.4°C between 1860–1920 A.D.
- Consistent trend of rise of temperature of mean value of 0.4°C between 1921 and 1945.
- Oscillating trend of rise of temperature between 1946 and 1975 A.D. around 0.4°C.
- Overall warming trend continued between 1976 and 1989 A.D. when temperature registered an upward rise by 0.26°C.
- The global rise of temperature reached its peak in the last decade of 20th century. Six out of seven warmest years on record since 1980.

A large number of temperature measurements indicate that the earth's average surface temperature rose about 0.75°C (about 1.3°F) compared with early twentieth century. The increase has not been smooth. It warmed noticeably from about 1910 to the mid 1940s, levelled off, and began increasing again in later 1970s. The 1990s were warmest of all, and 1998 was the warmest year. The year 2001 was the second warmest with an average temperature of 14.3°C (57.8°F). About two-thirds of the 20th century's warming has occurred between 1975–2000.

Box 3.1
Some Facts about Climate Change

- The climochronology reveals the fact that climates have changed in geological past, are changing now and it will also change in future.
- There has been consistent trend of rise in temperature of mean value of 0.4°C between 1921 and 1945.
- The global rise of temperature reached its peak in the last two decades of 20th century. The 1990s were warmest of all and 1998 was the warmest year.
- The long-term carbon dicxide equivalent greenhouse gases in the atmosphere have crossed 455 parts per million.
- During 1990s the carbon emissions grew by less than 1 per cent per year. But since 2000, the rate of emission has reached 3.5 per cent per year.
- There has been a marked toposphoric depletion of ozone gases and increase in greenhouse gases near earth's surface.
- Atmospheric green house gases continue to increase as a result of human activities.
- Decrease of about 10 per cent in snow cover has occurred since late 1960s.
- The average annual sea level rise throughout the 20th century was 0.17 mm and between 1962 and 2003 it rose to 1.8 mm.
- The edge of the West Antarctic ice sheet is shrinking at 400 feet every year.
- Ozone in troposphere does not absorb radiation but acts as a green house gas emission of nitrous oxide by aircraft engines which has increased ozone in the troposphere and has contributed to global warming.

Global Warming

As described above, temperature variations and climatic change has occurred during geological periods of the earth's history. Why, then is so much attention being paid now to the phenomenon of global warming if it is something that has happened before? This is because some of the warming that is taking place now is believed to have a different cause from previous warmings; it is what we call 'anthropogenic', which means it has been caused primarily by human activity, unlike previous global temperature shifts which were entirely due to natural cause. The scientific consensus is that human activities have been warming the world and it will continue to get hotter for the next hundred years or so, irrespective of whatever we choose to do now. In general, the new term, global warming, is used to explain the nature of climate change.

'The term "global warming" refers both to (a) the anthropogenic increase in the temperature that has already taken place on Earth's surface and near-surface air and oceans, and (b) its continuation into the future' (Susan M. Hodgson and David Philips, 2011). The *Oxford Dictionary of Geography* states that global warming is the 'increase in global temperatures brought about by increased emission of "green house" gases into the atmosphere'. In fact, global warming is a process by which the average temperature of the atmosphere, oceans and earth's surface is increasing. The characteristic features of global warming are:

(i) The global mean surface temperature has increased by a substantial margin of 0.6°C since pre-industrial times.

(ii) The concentration of greenhouse gases in the atmosphere has been significantly increasing at the rate between 0.25 and 5.0 per cent per annum.

(iii) The global temperature has risen in post two decades faster than any other 20-year period.

(iv) The 1990–1999 warm phase of the El Nino/Southern Oscillaton (ENSO) was unusual in context of past 120 years.

(v) The frequency as well as intensity of external weather events has become more variable.

(vi) Magnitude and pattern of climate change and variability are not uniform throughout the world.

(vii) There has been a marked pospheric depletion of ozone gases and increase in greenhouse gases near the earth's surface.

(viii) The frequency and intensity of extreme weather events have an increasing trend.

(ix) The diurnal range of temperature has increased.

The IPCC, the Intergovernmental Panel on Climate Change, was formed in 1988 by the World Meteorological Organization and the United Nations. Hundreds of scientists from 178 member nations do exceedingly careful reviews of climate data, and periodically report on their finding. A 1995 report said that there is a 'discernible human influence' on global warming. Then, in 2001, after 3 years of further study, an IPCC report went further and attributed most warming of the past 50 years to human activities. Several of their conclusions are as follows:

- Atmospheric greenhouse gases continue to increase as a result of human activities.
- The fraction of warming due to carbon dioxide exceed 50 per cent.
- Decrease of about 10 per cent in snow cover has occurred since the late 1960s.
- Reduction in the year's duration of lakes and rivers to ice cover in the northern hemisphere's mid- and high latitudes occurred in the twentieth century.
- Increase in the heat contained in the world's oceans have occurred since the late 1950s.
- Between 0.1 and 0.2 m rise in global average sea levels occurred in the twentieth century.

IPCC scientists did not believe that variance in solar radiation, natural climate fluctuation, or poor climate models explained the warming. As IPCC Chairman Rober T. Watson said, 'We see changes in climate. We believe we humans are involved, and we're projecting future climate changes that will be much more significant in the next 100 years than in the past 100 years'. The panel projects that, if current trends continue, increases in atmospheric greenhouse gases in the twenty-first century will result in an average global temperature increase between 1.4 and 5.8°C (2.5 to 10.4°F). About 75 per cent of the warming is expected to be specifically due to carbon dioxide. They also project sea-level rises of another 0.1 to 0.9 m. A sober forecast is that climate change will persist for many centuries.

Causes of Global Warming

A number of causes are responsible for the global warming. Apart from natural phenomena, the main causes are the 'anthropogenic', i.e. caused by human activities. Industrialization, urbanization, population explosion, shrinking agricultural land, etc. have had their contribution in global warming, but the main cause is the greenhouse effect or concentration of greenhouse gases. Therefore, it is necessary to explain the process of greenhouse and its effect on climate change/global warming.

Greenhouse Effect and Global Warming

The phenomenon commonly known as 'greenhouse effect' occurs due to emission of certain gaseous pollutants in the air which after the heating of the atmosphere causes the average global temperature to rise. The effect is produced by certain gases, such as carbon dioxide or water vapour, that cause a warming of the earth's atmosphere by absorption of infrared radiation. The atmosphere protects us by serving as a light scattering and heat mediating blanket. A large portion of heat emitted from the earth is reabsorbed by the atmosphere and is, in effect, conserved, with the result that the surface of the earth is warmer than it otherwise would be. In simple words, this warming is called the greenhouse effect, by analogy with the ability of a greenhouse to keep its inside warmer than the outside during the day time (Figure 3.1).

The atmospheric concentration of these gases has relatively increased since the advent of the industrial era. There has emerged a growing consensus in the scientific community that the anthropogenic greenhouse gas emission of carbon dioxide, nitrous

Figure 3.1 *Greenhouse Effect*

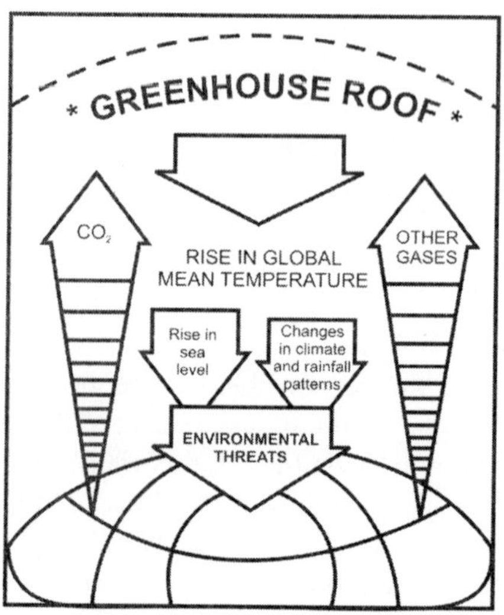

oxide, CFCs, methane and low level ozone threatens to disrupt human societies and natural ecosystem.

Recently, the world has been getting hotter. Over the same period the composition of the earth's atmosphere has been changed by increased human activity that produces CO_2 particularly by burning fossil fuels (in power stations, cars, lorries, ships and aeroplanes, etc.), cement production and deforestation.

These human-made CO_2 emissions account for 72 per cent of the added greenhouse effect to date. Once CO_2 is in the atmosphere it stays there for thousands of years before dispersing. Indeed, the world's most prestigious and influential body, researching global warming, the Intergovernmental Panel on Climate Change (IPCC) (2007: 17), warns that:

> ... both past and future anthropogenic carbon dioxide emissions will continue to contribute to warming and sea level rise for more than a millennium, due to the time scales required for removal of this gas from atmosphere.

The IPCC (2007: 10) also tells us that:

> Most of observed increase in global average temperature since the mid-20th century is very likely due to observed increase in anthropogenic greenhouse gas concentrations.

Per Capita Greenhouse Index

National ranking reflects the contributions of individual countries to the atmosphere's growing store of greenhouse gases. But it is also of interest to compare countries by per capita contributions of these gases, since the citizens of our planet vary widely in their relative responsibility for these emission.

Per capita figures also can be looked at as an indication of how much greenhouse gas emissions and global warming potential might grow if developing countries, in the process of development, significantly increase their per capita emissions. If just China and India, for example, raised their per capita emissions to the current world average, 1.1 metric tonnes of carbon dioxide heating equivalent per annum, total worldwide additions to the atmosphere would increase 28 per cent; if raised per capita emissions to the level of France, worldwide additions to the atmosphere would increase 68 per cent.

The increase and atmospheric concentration of greenhouse gases are responsible for climatic change, especially in rise of temperature, in other words, responsible for global warming. Changes in greenhouse gas concentrations are closely correlated with changes in the earth's surface temperature. It has been established that warming of global climate and rise in CO_2 and CH_4 levels coincide. Some evidence exists that predicted warming has begun to manifest itself already, but this evidence is controversial. Temperature records taken around the world since 1860 suggest that a global warming of 0.5° – 0.7°C has occurred since this year, with nearly half this rise occurring since 1965.

Among the most dramatic consequences of the greenhouse climatic disruption is an unprecedented rise in global sea levels to the extent of 8 to 25.5 inches by the end of the century, if gas emissions continue at the present rate. Melting of polar ice and glaciers, thermal expansion of seas would cause worldwide coastal flooding, and rise in sea level will result in submergence of many coastal areas. It will not only effect the ecological balance but will be a threat to the people and economy of the countries like Bangladesh, Netherlands, as well as to many coastal and island countries. In fact, the impact of worldwide climatic disruption could have hazardous consequences for the survival of the planet, both biologically and economically.

Effect of Global Warming

It is clear that the earth has been getting hotter. The scientists all over the world have examined the effects of global warming, even if the temperature rise is only few degrees. According to the IPCC (2007: 5) –

> ... warming of the climate system is unequivocal as is now obvious from observation of increases in global average air and ocean temperatures, widespread melting of snow and ice, and rising global average sea level. It would seem plausible that if human activities have added to global warming then reversing these actions should, in a few years time, lead to its reverse: global cooling. Unfortunately this will not happen. We noted above that once CO_2 gets into the atmosphere it stays there for a considerable time. Carbon dioxide is, of course, absorbed by vegetation, especially forest, and by water, for example, seas. The issue is that the increase in GHGs being emitted into the atmosphere now outstrips the capacity of forest and seas to absorb them. The consequences of this are truly awesome. ... the climate change

that takes place due to increase in carbon dioxide concentration is largely irreversible ... atmospheric temperatures do not drop significantly for at least 1000 years. (Solomon et al., 2008)

The effects of the global warming will be as follows:

Rising Sea Levels

A significant proportion (up to 80%) of the temperature rise is absorbed by the oceans, expanding the water and contributing to increase in the sea level. Between 1880 and 2000 it rose by 200 mm. The average annual sea level rise throughout the 20th century was 0.17 mm; between 1962 and 2003 it rose to 1.8 mm; and between 1993 and 2003 it reached 3.1 mm.

There has also been widespread melting of glaciers and ice caps particularly in Greenland and Antarctica, which has also contributed to sea level rise. Since 1978, Arctic sea ice shrunk by 2.7 per cent per decade and the Arctic ice cap is now less than half the size it was 50 years ago. The situation in Greenland is perhaps even more troubling; although its contribution to the rise in sea level has so far been less than that of Antarctica, observations in 2008–09 have identified rapid melting in the Greenland glaciers and fears have been expressed of rapid acceleration leading to a 'tipping point' (Houghton, 2009: 177–79). If the whole Greenland ice sheet melted, this would trigger a globally catastrophic rise in sea levels by 7 metres (23 feet). Even if these specific catastrophes do not occur, sea levels are still going to rise by at least 18–38 cm by 2100 even on the least pessimistic of IPCC's scenarios – and up to 50 cm on the most pessimistic scenarios.

Change in Rainfall Patterns

Long-term trends in precipitation (i.e. rain, sleet, snow, etc.) are much more variable than those in temperature. There have been increases in rainfall in eastern areas of North and South America, northern Europe and northern and central Asia. On the other hand, drying has occurred in the Sahel, Mediterranean and parts of southern Asia. More intense and longer droughts have been observed since the 1970s, particularly in the tropics and subtropics. Frequency of heavy precipitation has increased over most land areas (IPCC, 2007: 5). The IPCC (2007:15) predicts a further 10–20 per cent reduction in precipitation in the Sahel by 2090.

Rainfall pattern would be altered, with some areas getting more rainfall and others suffering more droughts.

Coastal Flooding

Coastal flooding will increasingly occur with rising sea level. People, who are already poor and overcrowded, may be forced from homes in low-lying countries such as Bangladesh. Prosperous countries such as the Netherlands, long accustomed to fighting back the sea, may lose land too. A rising sea level can exert great damage even in less critically affected regions, e.g. beaches, beach front property, and may increase storm surges along coastlines. A higher sea level also means that salty water can infiltrate fresh groundwater in coastal areas, making it undrinkable. Such infiltration of ocean water can also destroy coastal ecosystem and the species they support. Inland wetlands and ecosystem are also at risk.

Increased Likelihood of Storms and Heat Waves

Storms of increased severity are expected. This happens because trapped heat energy drives atmospheric air circulation and oceanic water circulation. These circulations have been positive forces because they distribute heat energy more evenly around the world – making earth on the average a more moderate place to live in. However, hurricanes and typhoons feed on this warmth too, and as the energy in air and water increases, storms can be expected to grow to greater severity than in the past.

Another logical accompaniment of increased warming is more severe heat waves. Higher temperatures would worsen air quality – recall those photochemical smog increases on hot summer days.

Ocean Acidification and Bleaching of Coral Reefs

The oceans are significant absorbers of atmospheric CO_2. As the concentration of CO_2 increases in the atmosphere, so does the amount absorbed into the oceans increase and this leads to changes in the acidity of the water. Even though the changes in ocean chemistry appear small (a decrease from pH 8.2 in preindustrial times to pH 8.1 currently), the effects on marine life and ecosystems are huge, including the destruction of coral reefs and the reduced capacity for sea creatures to develop shells and skeletons. One-third of coral reefs now appears to have been severely damaged by warming sea.

Spread of Diseases

Disease currently restricted to existing hot regions may move into the newly warming regions. Malaria is one such disease, spread by mosquito vectors, infecting and killing millions each year in warm climates. As temperature increases in currently temperate regions, mosquitoes and other insects are expected to move in, spreading disease as they go. In addition, indigenous disease organisms previously killed by winter cold will be better able to survive milder winters.

Loss of Plankton due to Warming Seas

The enormous (900 mile long) Aleutian Islands ecosystem of orcas (killer whales), sea lions, sea otters, sea urchins, keep beds and fish populations, appears to have collapsed due to loss of planktons – leading to loss of sea lions, orcas eating many sea otters, sea urchins explosion, with loss of keep beds and their associated fish populations.

Species would Migrate Poleward

Perhaps the most severe disruption of the terrestrial carbon cycle could come as result of the rapidity of the expected global warming. One consequence of this warming will be a movement towards the poles by a range of many plant species. But will plants be able to migrate poleward as their ranges shift? Species migration is usually a rather slow affair; a rapid rise of 2° C in global temperatures in the next 100 years could well translate into a 500–1,000 kilometre range shift poleward – a distance that many species will not be able to travel. This may result in the gradual extinction of many forest communities and, perhaps, a concomitant decrease in carbon storage rates.

Clouds Behaviour

The behaviour of clouds in future greenhouse scenarios is also a major source of uncertainty in global climate models. Clouds play an important role in regulating the earth's energy budget. They routinely cover about half the earth and account for a good deal of the 30 per cent of incident sunlight that the earth reflects into spaces. But clouds do not merely reflect solar radiation from the earth; they also absorb some of the heat radiated by the earth and reflect it back to the surface below. In other words, they also exert a greenhouse effect of their own. Currently, averaged over the globe, the reflective property of clouds dominate their greenhouse effect and thus, they exert a substantial net cooling effect on the earth.

Both the reflective effect and the greenhouse effect of clouds are large compared with the greenhouse effect exercised by CO_2 and other trace gases in the atmosphere. Recent data from a satellite-borne experiment shows that, on a scale where the warming effect from CO_2 doubling is 1, the cloud warming effect is currently 7. Therefore, the result is a net cooling effect four times larger than the expected warming that would be caused by a doubling of CO_2.

Fertilization Effect of Higher CO_2 Levels

Anticipating how this balance will change is complicated by the many competing influences that could prevail under greenhouse conditions. For instance, photosynthesis rates are known to be sensitive to atmospheric CO_2 level. The effect known as CO_2 fertilization can cause an average 30 per cent increase in growth rates in many, but not all, species when CO_2 concentration is doubled. Recent data suggest that this effect is temperature-dependent and could be increased up to threefold by a 3°C rise in air surface temperatures.

Further, there is evidence that CO_2 enrichment helps some plants minimize water loss, presumably making them more drought-tolerant – an important consideration if the drier conditions predicted by climate models for many mid-continent areas come to pass. If all these favourable effects of greenhouse conditions were realized, a net increase in plant matter might increase the carbon uptake of the biosphere and act as a negative climate feedback, helping to alleviate further CO_2 build-up. Indeed, there is evidence that the annual carbon uptake in the Northern Hemisphere has increased during the growing season in recent years, perhaps indicating greater carbon storage.

This potential rise in methane production is particularly worrisome. Methane is 20–30 times more efficient than carbon dioxide at trapping heat, and its concentration in the atmosphere has been increasing more than 1 per cent per year. Within 50 years, it may be the most significant greenhouse gas. Large methane releases from a warmer tundra – or from other sources sensitive to warming, such as deep ocean deposits – could create a major climate feedback.

Biosphere Response

Predicting how living systems will respond to global warming and what climate feedbacks they will exercise may be just as difficult as solving the riddles of ocean dynamics or cloud behaviour and provide still another major source of uncertainty in climate modelling. Terrestrial ecosystem plays an important role in the global carbon cycle, storing about 100 billion metric tonnes of carbon each year in plant materials through the process of photosynthesis and releasing about the same amount through the processes of respiration and decay. This annual flux is equivalent to almost 30 per cent of all the carbon held in the atmosphere, so even minor changes in the balance between photosynthesis, respiration and decay could affect the level of atmospheric CO_2 quickly.

Although the overall effects of global warming are relatively clear (as noted above), it is not easy to be specific about the precise consequences or locations of the changes that are taking place. According to Lewis (2006: 1), 'this is because the Earth is an integrated system consisting of a network of sub-systems … . Each linked by various processes … in complex ways'.

He makes the important point that it is not just the complexity of the world's *natural* systems that renders it impossible to make robust predictions but it is also their interactions with global political and social systems. The European Union (EU) Commission is more specific than Lewis in its predictions for Europe, which include the following potential effects:

- *Agriculture and fisheries*: increased risks of crop failure through extreme weather; reduction in soil fertility through depletion of organic matter; threats to fisheries from severe effects on coast and marine ecosystems.
- *Energy*: hydroelectric power will be affected both negatively and positively; negatively by the impact of drought in southern Europe leading possibly to a 25 per cent reduction; and positively in northern Europe by melting glaciers leading to an increase.

- *Infrastructure*: it will come under severer threat, particularly in low-lying urban areas through flooding.
- *Tourism*: it will suffer in alpine regions through snow loss and in Mediterranean regions through excessive temperatures.
- *Water*: it as well as the flooding danger from too much water, the increased stress placed on water supplies from too little water, particularly for irrigation and drinking, will be a major issue in southern Europe with 'Europe's high water stress area expected to increase from 19 per cent today to 35 per cent by the 2070s' (CEC, 2009: 4–5).

In global warming, increased carbon dioxide and warmer temperature can have a positive effect too. Agricultural crops, up to an unknown point are expected to respond with increased growth as the concentration of the nutrient carbon dioxide increases. However, the regions suitable for growing particular crops and trees will change. But this minor positive impact does not in any way justify the climate change due to global warming.

Prevention: Reducing the Risk of Global Warming

International Agreements and Effects

International communities are well aware with the problem of global warming. In this direction, United Nations Organization is doing commendable work, but still it has a long way to go. Not only several international conferences, seminars and workshops have been organized but important agreements have also been signed for international cooperation in this direction. The important conferences, agreements and declarations are:

Box 3.2

Important Conferences/Agreements and Declarations

1. First World Climate Summit, 1979, Geneva (Switzerland)
2. Conference on Industries and Climate, 1980, Vienna (Australia)
3. Vienna Conventions, 1985
4. Agreement on Montreal Protocol, 1987
5. Toronto Summit, 1985
6. Formation of Intergovernmental Panel on Climate Change (IPCC), 1988

7. Second World Climate Summit, 1990
8. UN Conference on Environment and Development known as First Earth Summit or Rio Summit, 1992 held at Rio de Janeiro (Brazil)
9. Signed Climate Change Convention during First Earth Summit, which aimed at reduction of carbon emission and stabilization of emission at 1990 level by 2000 AD, but it could not be implemented.
10. First Climate Change Convention also known as Berlin Summit, 1995
11. Second Climate Change Convention held at Vienna, 1996
12. UN Second Earth Summit, New York, 1997
13. Third Climate Change Summit held at Kyoto city (Japan), 1998 in which Kyoto Agreement was signed regarding 5.2 per cent cut in the emission of carbon dioxide by the developed countries

Box 3.3
Other UN Climate Change Conferences

4th	Climate Change Conference, Buenos Aires, 12 November 2000
6th	Climate Change Conference, Hague, 4 September 2001
7th	Climate Change Conference, Marrakesh, 11 November 2001
8th	Climate Change Conference, Delhi, 2 November 2002
10th	Climate Change Conference, Buenos Aires, 24 December 2004
11th	Climate Change Conference, Montreal, 29 December 2005
13th	Climate Change Conference, Bali, 1 January 2008
14th	Climate Change Conference, Poznam, 1 January 2009
15th	Climate Change Conference, Copenhagen, 30 December 2009
16th	Climate Change Conference, Cancun, 4 January 2011
17th	Climate Change Conference, Durban, 4 January 2012
18th	Climate Change Conference, Doha, 2 December 2012

Apart from above-mentioned conferences, agreements and conventions, in all the international conferences related to environmental issues, the climate change and global warming has remained an important point of discussion. Regarding reduction of greenhouse gas emission, special mention of UN Framework Convention on Climate Change and Kyoto Protocol is necessary.

UN Framework Convention on Climate Change

In 1992, at the UN Earth Summit in Rio de Janeiro, many countries signed this convention. Its objective was to stabilize 'greenhouse gas concentrations ... at a level that would prevent dangerous anthropogenic interference with the climate system'. Anthropogenic interference means human interference. About 150 countries ratified the convention. By ratifying it, a country was agreeing to return its greenhouse gas emissions to 1990 levels by 2000. In the United States, then President Clinton developed a Climate Change Action Plan to lower greenhouse gas emissions. However, the steps of the plan were voluntary and US emissions, far from decreasing, continued to climb. The same was true of many other countries. This result indicated that stronger measures were needed to reduce greenhouse gas emission and a plan that clearly specified: (a) the exact reduction in greenhouse gas emissions; and (b) when these reductions were required.

The Kyoto Protocol

After strenuous negotiating sessions, the Kyoto Protocol (also under the UN Framework Convention in Climate Change) was adopted in late 1997. Under the protocol, industrialized countries agreed to reduce greenhouse gas emission to an average of 5.2 per cent below 1990 levels between the year 2008 and 2012. Nations signing the protocol then laboured further to resolve three controversial points: (1) Find acceptable means of accomplishing greenhouse gas reductions; (2) Agree on an accounting system that gave nations credit for their reduction; and (3) Agree on penalties for lack of full compliance with agreed-upon reductions. This third was left until later to resolve.

In July 2001, 1,700 diplomats representing 178 countries resolved the first two of these points. Meanwhile, the United States had become only an observer to the negotiation because a new president, George W. Bush, refused to consider ratifying this 'fatally flawed' document. This made matters difficult for the protocol because, for it to come into force, 55 countries (accounting for at least 55%) of the world's 1990 emissions had to ratify it. With the United States – accounting for 25 per cent of the world's emissions – standing aside, it was difficult to reach 55 per cent although as of 2003, this may still happen. Cne US objection was that the protocol does not require less-developed countries to reduce their emissions. They were omitted because it is industrialized countries that are largely responsible for current levels of atmospheric greenhouse gases. And, industrialized nations are the ones able to develop the technology to lead the way in making reduction.

Copenhagen Summit

The 15th meeting of the Conference of Parties (Cop 15) to UNs Framework Convention on Climate Change, held from 7th to the 18th of December 2009 at the Bella Centre in Copenhagen (Denmark) was attended by 150 global leaders. The conference tried to enhance emission reduction targets for the developed nations during the second commitment period of the Kyoto Protocol.

Measures and Policies

To reduce greenhouse gas emissions or even to slow their rate of growth on a global scale will require an extraordinary degree of political consensus. Any one country or even one region can prevent the build-up of greenhouse gases by itself, although leadership by individual countries will be important in achieving global consensus. To be broadly acceptable, ideally, prevention policies should confer other local benefits on the adopting countries in addition to the worldwide benefit of reducing the risk of global warming.

Many policies have been proposed and are under study. It seems clear, however, that any successful prevention strategy will include five key elements:

- Increasing the efficiency of energy production and use,
- Switching from carbon-intensive fuels such as coal to hydrogen-intensive fuels like natural gas, where possible,
- Encouraging the rapid development and use of solar and other carbon-free energy sources,

- Eliminating the production of most CFCs and developing the means to recapture those now in use, and
- Reducing the rate of deforestation.

Four approaches suggested to cut greenhouse gas emission and reduction of global warming are: mitigation, adaptation, geo-engineering and conservation.

Mitigation

The word 'mitigation' has come into common usage in relation to reducing GHGs, because it has several varying meaning. The most apt is, 'To moderate or reduce in strength (heat, cold, light, etc.), so as to make more bearable, to temper the severity of (a climate)' (*Oxford English Dictionary*).

Lewis (2006:2) summarizes mitigation as: 'The basic solution to climate change but earlier articulated that most fossil carbons must not get into the atmosphere'. Currently the only proven way to do this is to leave most fossil fuels in the ground. That means, no new oil fields, no new coal mines.

Governments are already responding to this challenge, for example, by returning to the construction of new nuclear power stations and supporting the expansion of areas such as biofuel production. Funding is being put into the development of technologies to capture and store CO_2 from its points of production into storage facilities underground. Industry is also responding by increased investment in alternate energy sources such as wind and aiming to produce more energy-efficient technologies. Individual people are being asked to cut down on energy usage too. Groups have emerged that aim to tackle climate change and related problems; for example, the Transitions Network (TN, 2009) aims to drastically reduce emissions through re-localization and to address the depletion of fossil fuels through increasing social resilience.

Adaptation

The EU Commission's impact assessment identified three major options for adaptation:

- 'grey' infrastructure approaches based on engineering science including design of the built environment to withstand extreme events;

- 'green' structural approaches including increasing ecosystem resilience, halting biodiversity loss and restoring water cycles;
- 'soft' non-structural approaches including land use controls and economic incentives to reduce or prevent disaster vulnerability.

Central to these 'soft' approaches are water and land use policies to maintain ecosystems that are optimally resistant to climate change and minimize adverse effects of potential disasters particularly in relation to human well-being and health and the protection of property.

Geo-Engineering

Here the approach to reducing atmosphere CO_2 levels involves massive feat of engineering, some on a global scale. Possible examples, include removing gases from the atmosphere and solar radiation management. Several different proposals have been put forward for removing CO_2 from the atmosphere.

One is to find a way to remove it mechanically and store it either underground or in the oceans. Another is to find a way of removing it directly from the atmosphere, for example via 'synthetic trees'. Solar radiation management – reducing the amount of sunlight getting to the earth – can be undertaken chemically or mechanically. Chemically induced reduction in solar radiation takes place naturally, when sulphuric particles are emitted by volcanic eruptions. Mechanical approaches include the suggestion to use giant mirrors located in space to reflect the sun's rays away from the earth.

Conservation

The ultimate solution of global warming is that of conservation. Conservation of natural resources, along with substantial contraction of consumption, is an essential requisite of the deep environmental approach associated with strong notions of sustainability. This approach sees an intrinsic value in nature that must be sustained at all costs.

The following measures may be adopted by industrialists, individuals, as well as by the government for the conservation and reduction of GHGs.

Energy Efficiency

Increased energy efficiency appears to be the single and most promising option for reducing the risk of global warming. Higher efficiency in energy production and in a wide range of energy uses is the fastest and most cost-effective method for reducing emissions of carbon dioxide and other energy-related greenhouse gases. The replacement of a single 75-watt incandescent bulb with an 18-watt fluorescent bulb for example, provides the same light but prevents the emission of over 100 kilograms of carbon as CO_2 over the life of the new bulb. Replacing a car that burns 10 litres of gasoline per 100 kilometres with one that is twice as efficient would prevent the emission of more than five metric tonnes of carbon over a 10-year period. Experience during the past 15 years in the United States, Western Europe, Japan, China, Korea and many other countries has shown that large improvements in energy efficiency are possible without a detrimental effect on the rate of economic growth. The evidence suggests that economic growth is not tied inexorably to increased energy consumption, as was once thought. A recent study published by the World Resources Institute suggests that an energy strategy based on increased efficiency can sustain economic growth in both developing and industrialized countries.

More Efficient Power Generation

On the supply side, new power generation technologies offer significantly higher energy conversion efficiencies than have been available in the past. The spread of cogeneration offers some of the greatest potential savings. Cogeneration systems convert primary energy in the fuel to electricity and then use the otherwise-wasted heat to make steam or hot water for local processes requiring thermal energy. Such system can deliver between 60–80 per cent of the heat in the fuel to useful purposes, compared with about 44 per cent conventional steam electricity technologies. Cogeneration has many important applications in industry. In the US paper industry, for example, almost all steam process and heat requirements are met through the use of cogeneration systems. In Brazil, cogeneration systems in alcohol distilleries provide revenue from electricity sales as well as heat for distillation and purification processes. In Germany and Scandinavia, cogeneration systems are used widely to supply steam or hot water to heat nearby buildings and provide water.

Green Electricity or Carbon-free Energy Sources

Energy sources and fuels that produce no CO_2 are ideally suited to reduce the risk of global warming. To the extent that biomass energy is produced in a renewable manner – with growth of new biomass equalling that which is burned – no net CO_2 is added to the atmosphere. Such technologies are in operation already in many applications around the world, although in many cases they are not yet economically competitive with fossil fuels. Some carbon-free energy sources – such as hydroelectric dams and conventional nuclear power stations – raise serious environmental questions of their own. Nonetheless, hydroelectric, wind, solar, thermal, photovoltaic, and nuclear technologies deserve additional attention. As the eventual alternative to coal-based energy, carbon-free energy sources occupy a unique strategic role.

Green electricity is a carbon free energy source which can help in reduction of millions of tonnes of carbon dioxide, the main gas responsible for climate change every year.

'Green electricity' means electricity produced from sources which do not cause adverse impacts upon the environment. The solar energy, wind energy, hydro-power, waves, tital waves energy, geo-thermal energy, biomass-gas and landfill gas are included in green electricity category.

Solar Energy or Photovoltaic (PV) Energy: Solar energy is one of the most important sources of green electricity. A solar photovoltaic (PV) module works by converting sunlight directly into electricity (even on cloudy days) using semiconductor technology.

Solar energy can also be used to heat water directly using specially designed collectors. Even in winter a useful amount of hot water can be produced from roof top collectors. A third way to use solar energy is simply to design buildings to make maximum use of the sun. Using this so-called 'passive solar' approach, much of the energy that we currently use for heating, lighting and air conditioning can be saved.

Wind Energy

The wind power is one of the world's fastest growing energy sources! Wind turbine technology has greatly improved over the last ten years, making wind turbines quieter and more efficient so that electricity generated from the wind is now often competitive with traditional coal-fired and nuclear power stations. Wind turbines are also beginning to be built at sea – in the future much of our electricity could come from these offshore windfarms. Wind energy has been developed substantially in Netherlands, Denmark, France, Germany, USA, China, Spain, etc. In India wind energy is produced in states of Tamil Nadu and Gujarat and there is huge possibility of its development in many other states also.

Tidal Energy

Tidal power is generated at the occurrence of spring and neap tides. The turbines are installed in the estuaries of rivers and coastal areas. As the tide waves rise, the turbines start the generation of tidal energy. The coastal areas are suitable for the generations of tidal energy.

Waves Energy: Electricity can be generated with the help of strong ocean waves. A number of devices have been designed over the years to try and capture this huge energy resource.

Geothermal Energy

Geothermal energy comes from deep underground hot rocks . In some parts of the world steam comes to the surface and can be used to run steam turbines to produce electricity directly. In other places water can be pumped down and heated by the rocks to make steam. Geothermal energy can also be used to provide hot water and heating for buildings.

Biomass Energy

Either agricultural wastes or specially grown plants can be used as a fuel to run small power stations. As plants grow they absorb carbon dioxide (the main gas responsible for climate change) which is then released when the plants are burnt. So using biomass does not add any extra carbon dioxide into the atmosphere. Specially grown ‘energy crops’ provide not only an environmentally sound source of electricity, but also an important new opportunity for farmers. However, there are concerns about the sustainability of sourcing biomass from countries where forests are being cleared to make way for fast growing plants that are then used as biomass.

Landfill Gas

As rubbish decomposes in the landfill sites where our household waste is dumped, it gives off methane gas. This gas can be captured and burnt in a gas turbine to produce electricity. Burning the gas does give off carbon dioxide but since methane, which is emitted from the landfill site, is in fact a much more powerful greenhouse gas it is better to burn it than to allow the methane to escape into the atmosphere. There are already many landfill gas systems operating in the UK.

Hydropower is also considered as a source of green electricity. It is most common throughout the world and a main source of electricity generation.

Slowing Deforestation and Altering Agricultural Practices

Activities related to agriculture and forestry contribute significantly to greenhouse gas emissions. The largest emissions arise from land use changes, principally deforestation, which releases large quantities of CO_2 and methane to the atmosphere as well as N_2O and ozone. Therefore, reducing and ending deforestation ultimately is an immediate priority that would minimize emissions of two principal greenhouse gases dramatically. By suspending tax credits that encouraged deforestation as a means of establishing title to land in the Amazon region, Brazil has decreased the rate of deforestation sharply in the past two years, although the rate remains high.

Some agriculture practices release CO_2 and N_2O through energy inputs, fertilizer inputs, and soil erosion. Enclosing manure piles or feedlots would assist in the capture and use of methane as a commercial quality fuel, rather than allowing it to become an additional atmospheric burden. New strains of rich may not require the flooding now needed to control growth thus reducing methane production significantly. Revising agricultural practices to reduce requirement would reduce N_2O emission till agriculture can stabilize carbon in the soil rather than releasing it to the atmosphere through erosion, and improved productivity reduces pressure for deforestation, which currently emanates from the need to rotate continually among poorly performing agricultural plots.

Box 3.4

Steps Towards Control of Greenhouse Gases

Carbon Budget

In 1997 Greenpeace put forward the idea of a 'carbon budget' – the amount of fossil fuels that could safely be burnt in order to stabilize world climate.

Carbon Debt

A carbon debt is defined as the amount of time required for biofuel carbon offsets to repay the carbon debt created from clearing and burning tropical vegetation to plant biofuel crops. The debt can only be repaid if the production and combustion of the biofuels have net GHG emissions less than the life cycle emission of the fossil fuels that were replaced.

Carbon Tax

A tax on the use of fossil fuels, proposed in 1992 by the European Commission, aimed at reducing consumption of fossil fuels, to lessen the output of carbon dioxide into the atmosphere and thus counter the dangers of climate change. Denmark, Netherlands and United Kingdom have introduced such carbon tax.

Emission Credits/Carbon Credits

Carbon credits is a permit that allows the holder to emit limited carbon dioxide. Credits are awarded to countries or groups that have reduced their greenhouse gases below their emission quota. Through photosynthesis green plants remove carbon dioxide from the atmosphere. Kyoto Protocol proposed that if nations increase their forested areas, they be granted emission credits to count towards their CO_2 emission standards.

Carbon Footprints

A carbon footprint has been defined as the total set of greenhouse gas (GHG) emission caused by an organization or by an individual or a nation. This is measured by undertaking a GHG emission assessment denoted as carbon accounting.

Ecological Economics

A branch of economics which attempts to establish rules for possible sustainable economics, so that each economy is integrated into an ecological framework.

Ecological Footprints

The area of land functionally required to support a human society and society may easily exceed the carrying capacity of its own territory.

Carbon Managements

The monitoring of carbon emission and their removal through carbon sequestration.

Green Electricity

The electricity produced from sources which do not cause adverse impact on environment. These energy sources produce no CO_2 and are ideally suited to reduce the risk of global warming.

Policies and Politics after Kyoto Protocol

The Kyoto Protocol of 1997 was the beginning of the development of international policy and concerted action on global warming. Its purpose was to firm up the commitments made by virtually all the world's countries in 1992 at the United Nations (UN) Earth Summit which pledged to stabilize GHGs at a level that would prevent dangerous anthropogenic interference with the climate system.

There were two major problems with the Kyoto Protocol. The first is that it was only a very small beginning; the goal was for industrialized nations, on average, to cut their annual GHG emission by 2012 to 5.2 per cent below their level in 1990. This works out at 29 per cent below what they would otherwise have been if they had continued rising.

The second problem is the continued reticence among many countries for actually doing anything about global warming. The Protocol required ratification by enough industrialized countries to represent 55 per cent of the CO_2 emissions of the developed world and it took full seven years after the signing of the Protocol, for it to be enacted (Kyoto Protocol Statutes of Ratification, 2006). Australia only ratified it in 2008 and Kazakhstan in 2009. Up to recent times, the least enthusiastic country, and thus the biggest stumbling block to the full implementation of the Kyoto Protocol, has been the US, which at the time of writing had still not ratified Kyoto. Even though it has less than 5 per cent of the world's population, the US is responsible for about 25 per cent of total global warming.

The progress in controlling CO_2 is not satisfactory. Not a single country in the world is on target to meet its contributions to keeping the increase in CO_2 emissions at a maximum of 2°C above preindustrial levels – every country in the world is going to overshoot.

Nevertheless, some countries, particularly Sweden and Germany, and to a lesser extent the UK, are making some progress. Sweden has an effective carbon tax with a significant bite, so much so that when it was introduced, income tax was cut by half to compensate. Between 1990 and 2005, its economy grew by 44 per cent but its GHG emission went down by 9 per cent. Germany leads the world in microgeneration of energy from renewable sources. It has 'feed-in tariffs' where householders and small businesses get paid for generation of their own energy and inputting it to the national grid. The price is highly subsidized and fixed for 20 years or so. Germany is far ahead of any other country in microgeneration.

Developing countries are caught in a dilemma in relation to global warming. It is clear that they are the countries which will suffer the most from the impacts of global warming trend already set in by the developed countries, but they are also being warned that any attempt on their part to join the club of developed nations by expanding their economies must be constrained in order to minimize carbon emission.

The UN Climate Summit in Copenhagen in December 2009 was anticipated to produce a step-change in how countries would address environmental concerns. The accord, drafted by the US, China, Brazil, India and South Africa, speaks only of reducing gas emission to limit additional temperature rise to 2°C above preindustrial levels. There was no agreement on how to achieve these limits in concrete terms.

Several conclusions can be drawn from the preceding analysis of the scientific assertions surrounding global warming:

- The threat posed by global climate change to human and natural system is unparalleled in recorded history and means that the community of nations making up the planet must confront the likelihood of rapid and significant global warming.
- Technological and policy decisions made by nations in the relatively near term will affect substantially the timing and severity of any global warming that occurs.

- The technologies needed to prevent, reduce, or mitigate substantially further human impacts on the atmospheric concentrations of the gases contributing to global warming exist, although they are not disseminated widely.

Ozone Depletion (Thinning Ozone Layer)

Ozone depletion now has become one of the major environmental problems of the world, which has a direct impact on climate change or global warming. Ozone is a form of oxygen with three atoms instead of the normal two, and forms a fragile shield scattered in the stratosphere absorbing the sun's ultraviolet radiation, away from the earth's surface. It forms a three millimetre thick layer, but if it disappears or thins then all terrestrial life will be annihilated. Near the earth's surface, ozone is an increasingly troublesome pollutant, but safely up at a height of 20 to 30 kilometres from the earth in the atmosphere, it is as important to life as oxygen itself.

The thinning of this layer has generated global concern. This is due to several chemical pollutions discharged by industries or produced through other chemicals reactions. Chlorofluorocarbons (CFCs) are substances that are inert, inexpensive, stable, non-inflammable, non-toxic, easy to store and produce. The population explosion is one of the causes of the increase of CFCs due to use of air conditioners, insulation, solvents and fire-killers. In 1986, three US agencies, viz. National Aeronautics and Space Administration (NASA), National Oceanic and Atmospheric Administration (NOAA) and Federal Aviation Administration (FAA) started reporting on ground and satellite-based data on ozone depletion in stratosphere by the use of chemicals like halon 1301, methyl, the chloroform, carbon tetrachloride, solvents like FC-113, aerosols, foams and refrigeration CFC-11, and air conditioning CFC-12 that are depleting the ozone layer 4, 5, 8, 12 per cent, respectively, on global scale, because of human activities. As for as CFC regional consumption, USA is on top consuming 29 per cent, Russia and Europe 14 per cent, while in China and India it is merely 2 per cent. The total ozone amount in the northern hemisphere, middle and high latitudes have decreased, and sudden and unexpected decreases in the amount of spring time in Antarctic zone was confirmed at 1 to 20 km in the stratosphere ozone layer.

The increase in ozone layer depletion will invite the lethal ultraviolet rays from the sun which will increase cancer, eye damage, injure plants and marine life and even reduce our immunity to disease. The UV-B damages the genetic material DNA and also causes skin cancer. The ultraviolet radiation even reduces the efficiency of the immune system, thus impairing the body's resistance. Experts are of the opinion that animal and plant world would face a grim challenge if radiations are not checked. Certain crops have been found vulnerable to such radiations. Under impacts of radiations proteins in plants, microbes and animals would be affected and such changes in spatial orientation would be critical in the formed part of the reaction centre of an enzyme. Near cities where population concentration is more, ozone could be formed near the earth's surface having disastrous effect on human health, crops and ecosystems. It would also increase the incidence of smog in areas of industrial concentration. Ozone has had its effect on the earth's climate by adding to the greenhouse effect.

The world has now become serious about the most dramatic evidence of ozone depletion, a 'hole' in the ozone layer which appears every spring over the Antarctica. This *Antarctic Ozone Hole* has grown over the last ten years. Till now, no clear explanation has been found to explain this phenomenon and scientific research in this field is on progress. But general opinion regarding the causes of this depletion is the CFC. The countries of the world now have started serious thinking about this problem. In fact, this is the duty of the developed countries, who have technology and infrastructure, to check the ozone depletion. Because the problem is not a problem of individual nations, therefore, combined efforts can only check the ozone depletion.

The concern about ozone protection began by the late 1970s and early 1980s. In 1978 the United States banned the principal CFCs for most uses; and in 1980 European Community countries placed a limit on production of CFCs and cut their use in aerosol products by 30 per cent. Sweden, Norway and Canada also placed limits on CFC use. In 1975, UNEP had also expressed serious concern about ozone depletion and also developed a World Plan of Action on the Ozone Layer. Some of the important international actions taken on ozone protection are listed below:

1977	UNEP coordinating committee on the ozone layer was established.
1985	Vienna Convention for the protection of ozone layer.
1987	Montreal Protocol on substance that depletes the ozone layer.
1988	Entry into force of the Vienna Convention.
1989	Entry into force of the Montreal Protocol.
1990	Montreal Protocol amended in London. The amendment required signatory governments to regulate consumption and production of CFCs.
January 1991	Establishment of interim multilateral fund to provide financial assistance to developing countries in order to facilitate control measures.
May 1992	112 states ratified the amendment to the Montreal Protocol.
August 1992	Entry into force of the amendment.
November 1992	At Copenhagen, parties agreed to accelerate, phase out schedule, to January 1996

Among above-mentioned efforts, the Montreal Protocol is the most notable.

The Montreal Protocol began the process of strengthening the ozone layer protection regime. Because of new findings, most nations accepted the idea of some controls. However, the extent and the timing of the controls were contentious issues. Leading states argued for a freeze, followed by a gradual reduction in the production of ozone-depleting substances over a period of 10 to 14 years, whereas the veto coalition wanted a production cap. The Toronto Group still pressed for a cut, but as a compromise its members suggested a 50 per cent cut. As late as April 1987, the EC insisted it would agree to a reduction of no more than 20 per cent, but at the Montreal Conference, later that year, the EC representatives agreed to 50 per cent reduction.

Developing countries successfully pressed for some concessions; they obtained a derogation that would allow them to increase their CFC use for the first decade of the agreement. They were also to receive technical assistance as well as information and advice on CFC recycling and conservation. The fear was that without some concessions, Third World countries would set up their own CFC production plants.

But these concessions did not address all of the developing countries' major concerns regarding the evolving regime. Of particular concern was the fact that there were no provisions for a fund to help them make the switch to CFC substitutes. The fund was not established because the United States, Japan, and EC countries opposed it. Because there was no provision for financial assistance, three major developing countries – China, India and Brazil – refused to sign the protocol. These were important holdouts, because they had the potential to become major producers and consumers of CFCs. They refused to participate in the regime until the industrialized countries agreed to provide financial and technological assistance. UNEP's Tolba supported them in their calls for the establishment of the fund.

Scientific revelations about ozone depletion continued to unfold, and they spurred the regime strengthening process. The Montreal Protocol came into force on 1 January 1989 and the negotiations on its amendment began promptly. At a meeting to amend the Protocol in Helsinki in May 1989, the veto coalition began to shift its position dramatically. The EC members were among 80 nations voting for a complete CFCs phase out by the year 2000, but the same has not happened.

Ozone levels in the atmosphere have decreased about 50 per cent near the North Pole and the South Pole, about 4 per cent over Australia and New Zealand and about 3 per cent over the subcontinent of India. The amount of depletion varies with latitude (and with the season) because of variations in the intensity of sunlight.

This decline in ozone alarms scientists because stratospheric ozone intercepts some of the high-energy ultraviolet radiation coming from the sun. Ultraviolet radiation injures living cells by breaking strands of DNA and unfolding protein molecules. Species normally exposed to sunlight have evolved defense against average amount of ultraviolet radiation, but increased amounts could overwhelm those defenses. Land plants such as soybeans and rice would be subjected to sunburn and their yields would decrease. Even planktons in the uppermost 2 metres of ocean would be affected. Present research indicates an alarming decrease in phytoplankton primary productivity between 6 to 12 per cent in the coastal waters around Antarctica.

MULTIPLE-CHOICE QUESTIONS

1. What is climate?

   (a) Current atmosphere conditions

   (b) A summary of mean weather conditions over a time period

   (c) Changing pattern of weather conditions

   (d) All of these

2. Consider these statements

   1. Climate change refers to a large-scale change in one or more basic climate component.
   2. Climate change describes change in the average state of atmosphere over a time-scale ranging from decades to million years
   3. Climatic change has been occurring only during the 20th century.
   4. Variability, in both time and space, is an inherent feature of climate

   Which of the above statements (s) is are correct about climate change?

   (a) 1 and 3 (b) 1, 2 and 3

   (c) 2, 3 and 4 (d) 1, 2 and 4

3. According to IPCC the main cause of climatic change in modern time is/are?

   (a) Natural events (b) Anthropogenic

   (c) Volcanic activity (d) External forces

4. Global warming is an indication of:

   (a) Climate change
   (b) Desertification
   (c) Warming of the upper layer of atmosphere
   (d) Warming of hydrosphere

5. Consider the following statements:

   1. A warming trend in temperature was observed during the early 20th century that peaked around 1940.
   2. A somewhat lowering of temperatures was observed after 1940 which is attributed to global dimming.

   Which of the above statements(s) is/are correct?

   (a) Only 1 (b) Only 2
   (c) Both 1 and 2 (d) Neither 1 nor 2

6. Consider the following statements:

   1. Fossils found in sediments provide proxy evidence about ancient climates and climate change
   2. They are especially suitable sources of information for periods before Cambrian Era

   Which of the above statement(s) is/are true?

   (a) Only 1 (b) Only 2
   (c) Both 1 and 2 (d) Neither 1 nor 2

7. Which of the following statement(s) regarding earth's climate is/are true?

   (a) Earth's climate is changing as a result of natural and human processes
   (b) Earth's climate has remained stable over the history of the planet
   (c) Earth's climate will stabilize over the next century
   (d) Earth's climate has changed only once in geological periods

8. The International Panel on Climate Change (IPCC)

(a) Constructed the Kyoto Protocol

(b) Performed research in the field of climate change

(c) an international panel that has influence on biomes and economies.

(d) fines organizations that pollute

9. Match the following pairs:

A.	First World Climate Summit, Geneva	–	1. 1987
B.	Montreal Protocol	–	2. 1998
C.	Formation of IPCC	–	3. 1979
D.	Kyoto Protocol	–	4. 1988

Codes:

	A	**B**	**C**	**D**
(a)	2	3	1	4
(b)	3	1	4	2
(c)	1	2	4	3
(d)	4	3	2	1

10. In which Conference/Summit agreement 5.2 per cent on reduction in carbon emission was signed?

(a) Toronto Summit, 1988

(b) First Earth Summit, 1992

(c) Second Earth Summit, 1997

(d) Third Climate Change Summit, 1998

11. Which of the following is not a primary contribute to the greenhouse effect?

(a) Carbon dioxide

(b) Carbon mono oxide

(c) Argon

(d) Methane gas

12. Which of the following gases has an important role in maintaining atmospheric temperature?

(a) Nitrogen (b) Oxygen
(c) Argon (d) Carbon dioxide

13. In last 50 years, since 1960, the increase in concentration of CO_2 in our environment is about:

(a) 10% (b) 20%
(c) 6 % (d) 14%

14. Which of the following greenhouse gases, concentrations have increased the most since 1950?

(a) Water vapour (b) Nitrous oxide
(c) Methane (d) Carbon dioxide

15. Which human activity is largely responsible for increase in CO_2?

(a) Construction of Dams (b) Burning of fossil fuels
(c) Acid rain (d) Ozone depletion

16. Which among the following greenhouse gasses has the highest radioactive forces?

(a) Chlorofluoro Carbons (b) Carbon dioxide
(c) Methane (d) Ozone

17. The jet aircrafts fly very easily and smoothly in the lower stratosphere. What could be the appropriate explanation?

1. There are no clouds or water vapour in the lower stratosphere
2. There are no vertical winds in the lower stratosphere

Which of the statement(s) given above is/are correct?

(a) Only 1 (b) Only 2
(c) Both 1 and 2 (d) Neither 1 nor 2

18. As carbon dioxide in the atmosphere has increased in such a manner, how much has the acidity in the ocean waters increased?

(a) 5% (b) 10%

(c) 30% (d) 50%

19. Why is Carbon Tax necessary?

(a) to lessen the output of carbon dioxide in the atmosphere

(b) to collect funds for research

(c) to control the use of electricity

(d) to safeguard adverse effect of carbon dioxide on human health

20. Which of the following statements is/are not true?

(a) Carbon credit is a permit that allows the holder to emit limited carbon dioxide

(b) Carbon Management is possible only through gas management in upper strata of the atmosphere

(c) Ecological economics is a way for sustainable economics

(d) Carbon footprints is total set of GHG emissions caused by a unit

21. Human activities in the recent past have caused the increased concentration of carbon dioxide in the atmosphere but a lot of it does not remain in the lower atmosphere because of:

1. its escape into the outer atmosphere
2. the photosynthesis by phytoplankton in the oceans.
3. the trapping of air in the polar ice caps

Which of the statements given above is/are correct?

(a) 1 and 2 (b) Only 2

(c) 2 and 3 (d) Only 3

22. Regarding carbon credits, which one of the following statements is not correct?

    (a) The carbon credit system was ratified in conjunction with the Kyoto Protocol.

    (b) Carbon credits are awarded to countries of group that have reduced greenhouse gases below their emission quota

    (c) The goal of the carbon credit system is to limit the increase of carbon dioxide emission

    (d) Carbon credit are traded at a price fined from time to time by United Nations Environment Program

23. The UN Framework Convention on Climate Change (UNFCCC) is an international treaty drawn at:

    (a) UN conference on the Human Environment Stockholm, 1972

    (b) UN conference on Environment and Development, Rio de Janeiro, 1992

    (c) World Summit on Sustainable Development, Johannesburg, 2002

    (d) UN Climate Change Conference, Copenhagen, 2002.

24. Which one of the following is not an effect of the global warming?

    (a) Change in rainfall pattern

    (b) Change in the direction of ocean currents

    (c) Increased likelihood of storms and heat waves

    (d) Change in cloud behaviour

25. With the rising sea level, coastal flooding will increasingly occur. Which of the following country(s) will be affected by such event?

    (a) Madagascar and New Zealand

    (b) Indonesia

    (c) Bangladesh and Netherlands

    (d) Sri Lanka and Pakistan

26. What is the cause of the bleaching of coral reefs?

(a) Water Pollution (b) Warming of sea
(c) Marine vegetation (d) All of these

27. Which of the following will be the effect of global warming on agriculture?

(a) lesser growth of crops
(b) crop regions will change
(c) crop production will reduce
(d) arid conditions may reduce crop areas

28. In which year was the first World Climate Summit held?

(a) 1975 (b) 1979
(c) 1980 (d) 1982

29. In December, 2013, where was the 18th Climate Change Conference held?

(a) Doha (b) Durban
(c) Cancun (d) Bali

30. In UN Earth Summit (1992) at which level most of the countries agreed to reduce their greenhouse gas emissions?

(a) 1970s level (b) 1980s level
(c) 1990s level (d) 1992s level

31. Consider the following statements:

1. There has been a greater increase in the proportion of methane in the atmosphere since industrial revolution than in case of carbon dioxide
2. It is methane, rather than carbon dioxide that has been primarily responsible for increased temperature from year 1950 to 2000.

Which of the above statement(s) is/are true?

(a) Only 1 (b) Only 2

(c) Both 1 and 2 (d) Neither 1 nor 2

32. Montreal Protocol was concerned with:

(a) Control of ozone depletion

(b) Control of global warming

(c) Control of global warming and ozone

(d) Control of biodiversity and global depletion warming

33. What was the plan prepared by President Clinton to reduce greenhouse gas emission?

(a) Greenhouse Reduction Plan

(b) Reduction of CO_2 Plan

(c) Climate Change Action Plan

(d) Greenhouse Gases Management Plan

34. Consider the following statements:

1. Kyoto Protocol was signed to cut the emission of carbon dioxide by developed countries
2. Higher Greenhouse gas concentration is the result of anthropogenic interference
3. Green revolution has resulted in more emissions of carbon dioxide

Which of the above statement(s) is/are correct?

(a) 1 and 3 (b) 2 and 3

(c) 1, 2 and 3 (d) 1 and 2

35. In Northern Europe, which year was considered as 'the year without summer?

(a) 1780 AD (b) 1816 AD

(c) 1840 AD (d) 1845 AD

36. Who suggested shifting of earth on its polar axis as the factor responsible for climatic change?

(a) George Simpson (b) T.C. Chamberlain

(c) Robert Hook (d) Miltin Milankovich

37. Consider the following statements:

1. Ozone is broken down when it reacts with nitrogen oxide in the stratosphere
2. Ozone is formed when nitrogen oxide reacts with hydro-carbons in presence of sunlight

Which of the above statement(s) is/are correct?

(a) Only 1 (b) 1 and 2

(c) Only 2 (d) Neither 1 nor 2

38. What will be the impact of warming of seas?

(a) Growth of fisheries

(b) Loss of Plankton

(c) Lesser growth of marine plants

(d) All the above

39. Consider the following statements:

1. Between 1980 and 2006, the number of climate disasters has quadrupled
2. Climate change will effect two-thirds of the arable land in Africa could be lost by 2025
3. Due to climatic change new regions of agriculture will emerge

Which of the above statement(s) is/are correct?

(a) Only 1 (b) 1 and 2

(c) 1, 2 and 3 (d) 2 and 3

40. When was India's first National Action Plan or Climate Change (NAPCC) iaunched?

(a) June, 2008 (b) April, 2005

(c) June, 2010 (d) April, 2011

41. Which of the following is a man-made substance and a main cause of ozone depletion?

(a) Nitrous Oxide (b) Chlorofluoro Carbons

(c) Sulphur dioxide (d) None of these

42. Which of the following events have had direct impact on both regional and world climate?

(a) Tropical Cyclones (b) Ocean Currents

(c) El Nina (d) Glaciers

43. In which atmospheric layer is ozone commonly found?

(a) Troposphere (b) Stratosphere

(c) Mesosphere (d) Thermosphere

44. What is the life supporting role of ozone?

(a) It helps in precipitation

(b) Absorb the sun's ultraviolet radiations

(c) It controls air temperature

(d) All of these

Note: Questions given below (Q.No. 45 to 49) are based on **Assertion (A)** and **Reason (R)**. Select your answers using following codes:

(a) Both **A** and **R** are true and **R** is the correct explanation of **A**

(b) Both **A** and **R** are true but **R** is not correct explanation of **A**

(c) **A** is true **R** is false.

(d) **A** is false **R** is true.

45. **Assertion (A):** Climate change refers to a large-scale change in one or more basic climate components, such as temperature or precipitation.

Reason (R): Climate change have been caused more by anthropogenic factors, resulting in global warming.

46. **Assertion (A):** Global rise of temperature reached its peak in the last decade of the 20th century.

Reason (R): There is a consistent rise of temperature during earth's history.

47. **Assertion (A):** Increase in global temperatures brought about increased emission of 'greenhouse gases' into the atmosphere.

Reason (R): The frequency and intensity of extreme weather events have contributed to global warming.

48. **Assertion (A):** Amongst the most dramatic consequences of the greenhouse climatic disruption is an unprecedented rise in global sea level.

Reason (R): The melting of glaciers is the main cause of the rise of sea level.

49. **Assertion (A):** UNO is doing commendable work in developing awareness about global warming and its consequences.

Reason (R): IPCC is an agency which controls climatic changes in the world.

50. Consider the following statements:

1. An increase in tropospheric ozone concentration has a warming effect on earth.
2. A decrease in stratospheric ozone concentration has a cooling effect on earth.

Which of the above statement(s) is/are correct?

(a) Only 1 (b) Only 2

(c) Both 1 and (d) Neither 1 nor 2

51. Consider the following statements:

1. The last ice age reached its maximum about 18,000 years ago.
2. Rapid melting of ice in Holocene period after the last ice age raised the sea level by about 100 metres.

Which of the above statement(s) is/are true?

(a) Only 1 (b) Only 2

(c) Both 1 and 2 (d) Neither 1 nor 2

52. Which of the following point(s) is/are not included in Kyoto Protocol?

(a) Acceptable limit of greenhouse reductions

(b) Accounting system to give credit for their reduction

(c) Concessions to special category of nations for greenhouse emissions

(d) Agree on penalties for lack of compliance

53. Which of the following is/are the approach suggested to cut greenhouse gas emissions?

(a) Mitigation (b) Adaptation

(c) Geo-engineering (d) All of these

54. Which one of the following is included as a mitigation technique for GHGs?

(a) Construction of new nuclear power station

(b) Construction of new thermal power station

(c) Development of new oil fields

(d) Development of new coal fields

55. Consider the following statements:

1. Increased energy efficiency is the most promising option for reducing the risk of global warming
2. Conservation of natural resources along with contraction of consumption is one of the solution of global warming
3. Use of fossil fuels be stopped fully to minimize the global warming

Which of the above statement(s) is/are correct approach of reducing the global warming?

(a) Only 1 (b) Only 3

(c) 1 and 3 (d) 1 and 2

56. Recent evidence from polar ice cone analysis shows that carbon dioxide levels have never exceeded ... ppm in the last several hundred thousand years.

(a) 500 (b) 1000

(c) 25 (d) 300

57. What could be the reason for the formation of ozone hole over the Antarctica region?

(a) Presence of prominent polar front and stratospheric clouds and inflow of chlorofluoro carbons.

(b) Temperatures increase at polar region.

(c) Inflow of methane and chlorofluoro carbons.

(d) Tropospheric turbulence and inflow of chlorofluoro carbons.

58. Rice cultivation in many regions may be contributing to global warming. What is the possible cause of it?

(a) The anaerobic conditions associated with rice cultivation cause the emission of methane

(b) Where Nitrogen based facilities are used, nitrous oxide is emitted from the cultivated soil

Which of the statements given above is/are correct?

(a) Only 1 (b) Only 2

(c) Both 1 and 2 (d) Neither 1 no r2

59. Concept of Carbon Credit originated from which of the following?

(a) Earth Summit, Rio de Janeiro

(b) Kyoto Protocol

(c) Montreal Protocol

(d) G-8 Summit, Heiligendamm

60. During 2009, UN climate summit in Copenhagen, USA, China, Brazil, India and South Africa drafted an accord for reducing gas emission. What was the limit fixed for it?

(a) to limit gas emission below pre-industrial levels.

(b) to limit additional rise to 2°C above pre-industrial limit.

(c) no limit was fixed

(d) rise in temperature be leveled by technological means.

61. Which one of the following country is the first country in the world to propose a carbon tax for its people to address global warming?

(a) Japan (b) Australia

(c) New Zealand (d) Germany

62. What will happen if the temperature of lower atmosphere rise further?

1. People living in small Islands will suffer most
2. Coral reefs will have a higher biotic productivity and flourish
3. Agricultural production in middle and higher latitudes will increase
4. Incidence of floods and draughts will increase

Which of the above statement(s) is are correct?

(a) 1, 2 and 3 (b) 2, 3 and 4

(c) 1, 2 and 4 (d) 1, 3 and 4

63. Land use changes are considered the second most important cause of increasing carbon dioxide fossil fuels. Which land use practices is/are responsible for it?

(a) Increasing multiple cropping

(b) Irrigated paddy farming

(c) Deforestation for agricultural and other development purposes

(d) Increasing productivity through intensive farming

64. Which of the following is the cause of the variations in ozone depletion?

(a) Varies according to forest cover area

(b) Varies with latitude and seasons

(c) Varies with agricultural land use

(d) Varies in various temperature belts

65. Which of the following is not an impact of ultraviolet radiation?

(a) Injures living things by breaking strands of DNA

(b) Reduce immunity to disease

(c) Damage microorganisms of the soil

(d) Injures plants and marine life

66. Green electricity means:

(a) Electricity produced by trees

(b) Thermal electricity

(c) Electricity produced by those sources which are not harmful to environment

(d) Low cost production of electricity

67. Which one of the following is not a source of green electricity?

(a) Wind image (b) Biomass energy

(c) Geothermal energy (d) Thermal Energy

Answers

1. b	2. d	3. b	4. a	5. c	6. b	7. a
8. c	9. b	10. d	11. c	12. d	13. d	14. c
15. b	16. c	17. c	18. c	19. a	20. b	21. c
22. d	23. b	24. b	25. c	26. b	27. b	28. b
29. a	30. c	31. a	32. a	33. c	34. d	35. b
36. c	37. b	38. b	39. c	40. a	41. b	42. c
43. b	44. b	45. a	46. c	47. c	48. b	49. c
50. c	51. c	52. c	53. d	54. a	55. d	56. d
57. a	58. a	59. b	60. b	61. c	62. d	63. c
64. b	65. c	66. c	67. d			

4 Sustainable Development

Man has continually, and at an increasing rate, been changing the forms and modes of his interaction with the environment. Natural processes and factors that depend on the structure and other characters of the earth, and social processes and factors that depend on structure and characteristic of our society, have been closely interwoven. On the one hand, population of the earth is growing rapidly and now it has reached the 7.5 billion mark, while on the other hand, with the technological and scientific revolution, developmental activities have been accelerated to such an extent that in certain areas they have become a threat to the environment. The consumption of natural resources is increasing at an alarming rate, causing great concern for their conservation. The degradation of the environment, due to industrial and other wastes discharged into the atmosphere and hydrosphere, has also caused great concern and there is a big question – what will happen next? All these environmental problems associated with development have raised several questions regarding the type and nature of development and this has given rise to the concept of 'sustainable development'.

Concept of Sustainable Development

The term 'sustainable development' was first used in the Cocoyoc Declaration on Environment and Development in the early 1970s. Since then, it has become a buzzword for international organizations dedicated to achieving environmentally benign or beneficial development. The term has served to catalyse debate over the relationship between economic change and the natural resource base in which it is grounded. Definitions vary, but the most widely adopted one is contained in *Our Common Future,* 1987 report of the World Commission on Environment and Development: 'Sustainable development is the development that meets the needs of the present without compromising the ability of future generations to meet their own needs'. This definition is far from adequate – it begs the critical question of what constitutes an acceptable 'need' in the context of present global situation– but it does encapsulate the essential point – that we all have to learn to live within our planetary means. Various UN publications, in an effort to define strategy of sustainable living and development, refer to it as 'improving the quality of human life while living within the carrying capacity of the supporting ecosystem'.

The environment, society and economy are integrated in such a way that each element plays a vital role. The entire concept of sustainable development can be understood through Figure 4.1.

It becomes clear from the above figure that better quality of life in cities and communities can be developed through long-term environmental planning based on sustainable economic

Figure 4.1 *The Concept of Sustainable Development*

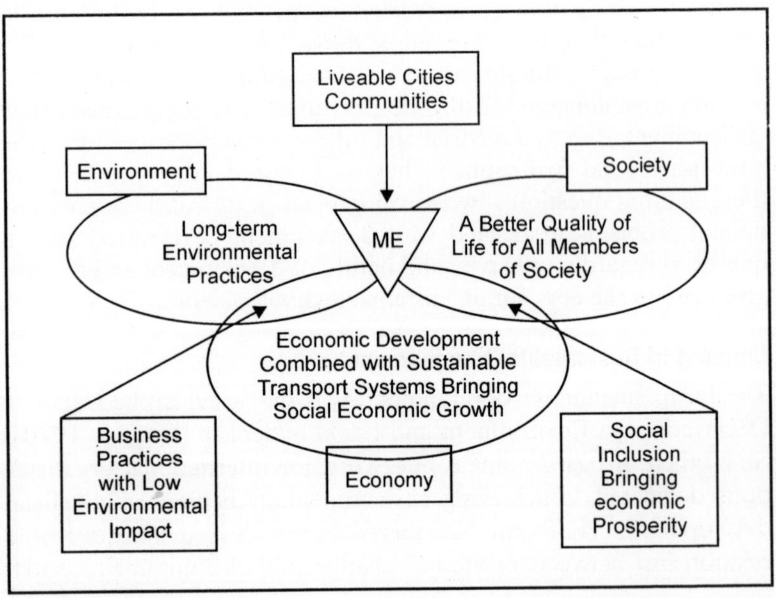

Box 4.1
Definitions of Sustainable Development

- A strategy that meets the needs of the present generation without compromising the ability of future generation to achieve their own requirements.
- Improving the quality of human life while living within the carrying capacity of supporting ecosystem.
- Environmental care 'married' to development.
- An environmental 'handrail' to guide development.
- Balancing the fulfilment of human needs with the protection of the natural environment.

development. The growth of economy should be directed towards the protection of environment and social well-being.

A matrix of sustainable development shown in Figure 4.2 indicates how planet Earth and its people engaged in various profit making activities can adopt an approach of sustainable development.

Figure 4.2 *Sustainable Development Matrix*

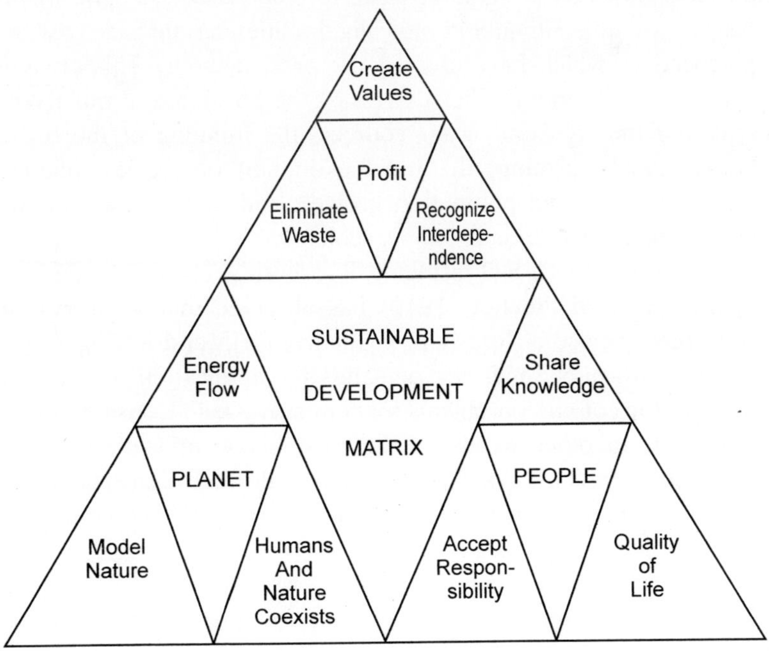

Growth of the Concept of Sustainable Development

The concept of sustainability has a long historical background. In ancient Indian writings, 'nature' or environment has been considered as the controller of all human activities, including economic development. They have given the status of God to all the components of environment, such as air, water, land, natural vegetation, animals, etc. In spite of the fact that there was no need for their conservation, the concept of their protection has been put forward in all the religions of India. In Greek writings, there is a reference of *'Ge'* or *'Gaia'* as the Goddess of the Earth – the mother figure of natural replenishment. In fact, that was a time when population was limited and there was mutual adjustment between man and environment.

It was only after the industrial revolution, followed by transport revolution and urbanization, that degradation of the environment started and which has given rise to the concept of sustainable development. The sudden interest in sustainable development is, of course, a response to warnings that the world is facing an environmental crisis and, therefore, social catastrophe in the near future unless mankind radically modifies certain practices and perspectives which have created the present crisis. This crisis is characterized, among other things, by the poisoning of our rivers, seas and underground water sources; the thinning of the ozone layer; global warming; the rapid extinction of species; massive deforestation; rapid population growth; and uncontrolled urbanization, with its attendant social problems.

The former US President, Roosevelt's chief forester and mentor, Gifford Pinchot (1910) has observed that 'conservation advocates the use of foresight, prudence, thrift and intelligence in dealing with public matters, demands the application of common sense to the common problems for common good'. Conservation of resources, in other words, is the 'sustainable utilization' of the resources, which can be achieved through environmental regulation, management, adequate surveillance and accountability. The environmental degradation and associated problems, like desertification, soil erosion, tropical forest depletion, etc. in the Third World, are the consequences of present development policies. Therefore, there is need for positive mutual adjustment between development and natural environment.

Several strategies, policies and concepts have been propounded by ecologist and planners to deal with environment/natural resources management, such as the zero growth strategy of Daly (1977), organic agriculture for self-sustaining societies of Naess (1977), a narrow and overspecialized approach of environmental economics by Norgaard (1985) and the well-known concept, i.e. use of environment-friendly techniques and products.

The publication of *US Global 2000 Report* (1980) and other publications like *Resourceful Earth* (Simon and Kohn, 1984), *The Global Possible* (Repetto, 1985), and *Our Common Future* (World Commission on Environment and Development, 1987) have generated a debate highlighting the consequences of overexploitation of natural resources and neglect of global 'common interest'. The various programmes of the UNO, specially the programmes of UNEP, have emphasized the need for sustainable development also referred to as 'eco-development'.

Despite its simplicity, the phrase 'sustainable development' is capable of carrying a wide range of meaning, sometimes allowing quite different ideas about what should be done to exist side by side without the conflict between them being clear. Both radical environmentalists and conventional development policy pragmatists have seized the phrase and used it to express and explain their ideas about development and environment. In the process they have created a theoretical maze of great complexity. The superficial conformity of writing and thinking about sustainable development hides very real and fierce battles behind the scenes over the meaning.

Sustainable Development of Rio, 1992

The United Nations Conference on Environment and Development (UNCED) at Rio in 1992 (the Earth Summit) was attended by 128 heads of states and in total by the representatives of some 178 governments. Debate at the conference drew very directly on the mainstream ideas about the environment and development that had evolved during the 1980s. The Rio Declaration and the much larger Agenda 21 were the fruits of endless negotiation at a series of Preparatory Commission meeting and at Rio itself, between teams of diplomats determined to surrender as little as possible of their national interest. A key feature of these debates was the gap between countries in the industrialized North and the underdeveloped South, which became steadily more glaring in the run up to, and during, the conference. The issues of climatic change and biodiversity that dominated the Rio conference are vitally important to certain countries (especially those vulnerable to sea-level rise), but they are not the principal environmental problems faced by most countries of the South.

The Rio Declaration ended up as a bland list of 27 principles. It was long-winded, and in places self-contradictory; even after a long debate, the US delegation released an 'interpretative statement' that effectively dissociated it from a number of the agreed principles. It dissented from principle 3 that there was a right to development, with the Americans arguing that development was not a right but simply a shared goal, and from any interpretation of principle 7 that suggested that there was an international liability to make development sustainable (i.e. for rich countries to pay for it). Choices about development and environment were matters for individual countries to make up their minds, and not an issue over which they should be subject to international opinion.

Agenda 21 was even more burdened by divergent opinions, becoming a bloated volume of more than 600 pages in 40 separate chapters. Its scope was enormous, covering issues from biodiversity and water quality to the role of women, children and organized labour in delivering sustainable development. It reflected previous mainstream thinking about sustainable development in several ways. First, it made the need for economic growth a central theme, as in the Brundtland Report. In the sustainable development mainstream, everything is predicted on economic growth, both globally and nationally. Second, Agenda 21 emphasized the familiar straightforward issues of environmental management; all the familiar environmental issues from the *World Conservation Strategy* appear developed but unmistakable. Third, Agenda 21 was technocentric. The first six key themes make this quite clear; growth with power and technology will direct the evolution of policy towards more efficient use of the environment and hence towards a more sustainable world economy. The essential means to achieve sustainability also reflect this technocentrism, building on information, science and environmentally-sound technology. Fourth, Agenda 21 presumed a multilateral benefit to be derived from a sustainable development strategy, as did the Brundtland Report. It suggested that change would arise from the mutual interest of industrialized and non-industrialized countries, and from the concern of present generations about the future. It suggested resources and technology to flow, directed and promoted by international agencies and structured and regulated by international legal instruments. Fifth, like its predecessors, Agenda 21 called for sustainable development through participation. As in *Caring for the Earth*, women, children, young people, indigenous people, trade unionists, businessmen, industrialists, farmers, local authorities and scientists are all summoned to play a role – a rainbow coalition to put flesh on the endless skeleton on the text of Agenda 21.

After Rio

At the close of Earth Summit +5 – the nineteenth special session of the United Nations General Assembly, held from 23 to 28 June 1997 in New York – delegates from over 165 countries adopted the *Programme for the Further Implementation of Agenda* 21. In the final document, government acknowledged that the global environment has continued to deteriorate since 1992, with rising levels of greenhouse gas emission, toxic pollution and solid waste. Renewable resources, notably fresh water, forest, top soil and marine fish stocks continue to be used at rates that are clearly unsustainable.

To address these concerns, governments took action on several fronts at the special session, as reflected in the final document. Among other decisions, they agreed to reconfirm the political commitment to sustainable development from all members of the international community, as well as from all major groups of civil society:

- Reconfirm the financial commitments;
- Establish an intergovernmental forum on forests under the Commission on Sustainable Development to continue policy dialogue on this issue, including more focused consideration of elements for a possible legal instrument;

- Open high-level, inter-governmental dialogues on fresh water, and on energy and transport, which will be taken up by the Commission on Sustainable Development at upcoming sessions;
- Make a stronger commitment at the global level to such issues as tourism, changing production and consumption patterns, and eco-efficiency; and
- Set a more focused work programme for the Commission on Sustainable Development.

Since Rio, there have been three further milestones. First, the Kyoto Protocol was negotiated in 1997, recognizing the importance of a global reduction in GHG emission in order to slow down the effect of climate change. Second was the Millennium Summit and Millennium Development Goals (MDGs). Third, in 2002, the Johannesburg Earth Summit was held. It looked back at the

Box 4.2
Brundtland Report, Agenda 21 and the MDGs

Brundtland Report (1987)	*Agenda 21 (1992)*	*MDGs (2000)*
Common challenges	Key social and economic dimension	Eradicate extreme poverty and hunger
Population and human resources	International cooperation	Primary universal education
Food security: sustaining and potential	Combating poverty	Promote gender equality and empower women
Species and ecosystem resources for development	Changing consumption patterns	Reduce child mortality
Energy: choices for environment and development	Demographic dynamics	Improve maternal health
Industry: producing more with less	Human health	Combat HIV/AIDS, malaria and other disease
The urban challenge	Human settlements	Ensure environmental sustainability
Common endeavours	Strengthening the role of major group	A global partnership for development
Managing the commons	Women	
Peace, security, development and the environment	Children and youth	
Towards common action: proposal for institutional and legal change	Indigenous people	
	Farmers	
	Implementation	
	Education and awareness	
	Capacity building in developing countries	
	International institutions	

progress of Agenda 21 and also set targets for sanitation, energy use, health, agriculture and biodiversity. In addition to meetings, reports and treaties as mentioned earlier, the Millennium Declaration and MDGs are of great significance to global environment policy. These were built on a number of UN global conferences during the 1990s. The Millennium Declaration marked a strong commitment to the right to development, to peace and security to gender equality, the eradication of many dimensions of poverty and to sustainable human development (UN, 2005: v)

Box 4.2 outlines the main components of the Brundtland Report, Agenda 21 and the MDGs. The similarities between these are striking.

The similarities in these documents indicate the close-knit relationship between international development and sustainable development. In all these documents, further commitment was asserted between developed and developing nations on sustainable development. There is a need for mutual understanding between developed and developing countries so that sustainable development can be possible, at least to a moderate limit, but this has still a long way to go.

Rio + 20

United Nations Conference on Sustainable Development, 2012 was held in Rio de Janeiro from 20–22 June 2012, 20 years after the first Rio meeting in 1992. Green economy and institutional framework for sustainable development were two focal themes along with poverty reduction.

The Rio+20 Declaration was entitled as 'The Future We Want'. Its main points were:

(i) Eradication of poverty is the greatest global challenge facing the world today and it has an indispensable requirement for sustainable aspects.

(ii) Need to mainstream sustainable development integrating economic, social and environmental aspects.

(iii) Promoting sustainable development by economic growth, economic opportunities for all, reducing inequalities, raising standard of living and integrated sustainable management of natural resources and ecosystem.

(iv) Reaffirming support to accelerate the achievement of MDGs by 2015.
(v) Combatting climate change requires urgent action according to UNFCCC.
(vi) Gender equality and women empowerment are important for sustainable development.
(vii) The contribution of NGOs have also been recognized and further promotion of their work is necessary for implementation of the policies of sustainable development.

Box 4.3
Green Economy and Green Agenda

Green Economy

Green Economy can be defined as one that results in improved human well-being and social equity, while significantly reducing environmental risks and ecological scarcities. Green Economy argues that the environmental cost of current development should be taken into account while measuring economic growth. Developed countries consider Green Economy as an opportunity that will generate jobs, increase their competitiveness and promote resource efficiency. The concept of Green Economy was popularized by UNEP in 2008.

Green Agenda and Brown Agenda

Green Agenda is concerned with 'inter-generational equity', i.e. equity between the present and future generations. Green agenda concerns are global and for the human well-being. On the other hand, Brown Agenda deals with environmental issues associated with urban and industrial areas.

Eco-development

The concept of eco-development implies that all developmental activities should be performed or designed in such a way that regional ecological balance can be maintained. This was originally advocated by UNEP for environmental planning and was defined as 'development at regional and local levels consistent with the potentials of the area involved, with attention given to the adequate and rational use of the natural resources, and to application of technological style' (UNEP, 1975).

Dasmam (1985) described eco-development as a triangle, one side of which is *basic needs*, the second *self-reliance*, and at the base is *ecological sustainability*. Each of these variables needs to be considered, since development will not be sustainable unless poor people are involved in meeting their aspirations. In fact, 'the environment and development are means, not ends in themselves. The environment and development are for the people, not people for environment and development' (Chamber, 1986). There is thus a need for 'environment thinking' with 'development thinking' and 'sustainable livelihood thinking'. In Third World countries, the main consideration should be given to 'basic needs' and 'self-reliance' while designing eco-development for a region; consideration should be given to both short-term and long-term planning. For sustainable eco-development, it is also necessary to give priority for the livelihood of the poor; this can be achieved through political change at the local, national and international levels. The political, economic and social forces should be directed to fulfil the needs of the population in such a way that environment remains protected.

Basic Aspects of Sustainability

The question of sustainable development has emanated from overexploitation of resources as well as mismanagement of technology. The aspects, which require monitoring of sustainability, include climatic change, biodiversity, disposal of hazardous and toxic wastes, disposal of pollution generating industries and food and ecological security.

Swaminathan (1991) has identified nine principles for success in promoting ecologically-sound agriculture. These are: (i) land, (ii) water, (iii) energy, (iv) nutrient supply, (v) genetic diversity, (vi) pest management, (vii) post-harvest system, (viii) systems approach, and (ix) location-specific research and development.

The deterioration in the ecological base in various countries in spatio-temporal terms due to irrational management of the resource and environmental systems having damaging repercussions is reflective of unsustainable policy frame and planning strategies followed so far. Its observable signs can be listed as:

(i) extensive deforestation accounting for loss of flora, fauna and some rare species;

(ii) drying up of drinking water resources and fall in the underground water levels;
(iii) intensifying rate and frequency of flood and droughts;
(iv) land degradation due to desertification, wastelands, salinity and waterlogging;
(v) deterioration in quality of air and water;
(vi) pressure of population resulting in unemployment and mass migrations; and
(vii) unplanned urbanization and unprecedented growth of urban slums, etc.

The environmental problems are multidimensional and varied in nature in both developed and developing countries. On the one hand, there are global problems which have impact throughout the world. On the other hand, every country has its own development as well as environmental problems. Apart from this, regional and local problems need immediate attention. The problems created by technology transfer from developed countries to Third World countries have become a cause for concern, because in the absence of proper management it has resulted in environmental degradation. Such alarms are embodied in the unsustainable activities which may include:

(i) intensive cultivation of land without taking adequate care of soil fertility;
(ii) development of irrigation facilities without proper water management, which leads to waterlogging, alkaline or saline soil;
(iii) improper use of pesticides, fungicides, herbicides, etc. which cause soil damage and biological imbalance;
(iv) excessive trapping of underground water accounts for steep fall in underground water levels;
(v) replacement of high yielding hybrid varieties lead to spread of disease capable of wiping out the entire crop as it happened with Irish potato crop in 1985 and Bengal rice famines in 1942;
(vi) excessive use of non-degradable material like plastic creates problems of waste management;
(vii) discharge of industrial and municipal waste in water bodies leads to problems like unmanageable water pollution; and
(viii) automobiles and industries have become a major cause of air pollution.

In fact, development without proper management has become a cause of eco-destruction, for which sustainable development is the only solution. In a report for a sustainable world, following steps have been suggested:

1. A production efficiency era of minimum environmental damage costs through *energy* transition.
2. For a stable world population, a *demographic* transition.
3. From non-renewable to renewable *resource* transition.
4. By a global mutually agreed objective between South and North, a *political* transition.

While all these initiatives are necessary, the critical shift required is in the perception of man. For 'industrial man' has grown accustomed to thinking that (a) the world's resources are infinite, and (b) they exist primarily to be exploited for the benefit of mankind. These assumptions are leading us to disaster. What is required is that we replace this outmoded perspective with one that sees us all as crew members of spaceship earth and fragile life-support system spinning through space.

Guiding Principles of Sustainability

The guiding principles of sustainability cut across ecological, economic, social and cultural dimensions and there are obvious trade-offs. What needs to be addressed is sustainable development – for whom in what context, and with what objectives?

The guiding principles of ecological sustainable resource management (Hare et al., 1990) are as under:

- Inter-generation equity providing for today while retaining resources and options for tomorrow.
- Conservation of cultural and biological diversity and ecological integrity.
- Constant natural capital and 'sustainable income'.
- Anticipatory and precautionary policy approach to resource use, erring on the side of caution.
- Use of resources in a manner that contributes to equity and social justice while avoiding social disruption.
- Limits on natural resource use within the capacity of the environment to supply renewable resource and assimilate wastes.
- Qualitative rather than quantitative development of human well-being.

- Pricing of environmental values and natural resources to cover full environmental and social costs.
- Global, rather than regional or national, perspective of environmental issues.
- Efficiency of resource use by all societies.
- Strong community participation in policy and practice, during the process of transition to an ecologically sustainable society.
- The importance of individual, local and institutional sustainability to agricultural and environmental sustainability needs to be stressed.
- Poorly conceived globalization/trade liberalization threatens the sustainability and existence, economically and practically, of many developing countries.

Temporal dimension of sustainability becomes obvious when we try to relate it to attributes that characterize ecosystem. A miniature ecosystem of microorganism created in culture flask may have a brief turnover time of less than an hour. On the other hand, a village pond, grassland or a forest ecosystem has a turnover time that may have to be viewed in terms of decades or hundreds of years. And, the time frame becomes even greater for a landscape of interacting ecosystem.

Measuring Sustainability

It is well recognized that conservation and sustainable development represent two sides of the same coin, closely interlinked with one another, as one cannot be achieved at the expense of the other. From a human angle, such an integrated approach demands satisfying basic human needs in an equitable manner, maintenance and promotion of social, cultural and biological diversity, and indeed the ecological integrity of the system as a whole. It is in this context that one needs to view the whole question of evaluating sustainable development.

Indicators for Evaluating Sustainability

Sustainability cannot be measured by a single indicator. A number of different indicators – ecological, economic and social – need to be reconciled in any evaluation of sustainable development of a landscape unit. The economic indicator is the most obvious one.

Energy efficiency is an ecological indicator, which is also helpful in the decision-making process.

How does one monitor sustainable development? We need a variety of indicators. The indicators of sustainable development are varied; therefore compromises are called for. Monitoring and evaluation have to be done using a number of diverse indicators that may be ecological, economic, social and even cultural. Basically, these indicators add up the 'value system' that a given society desires to have.

Ecological Indicators

Ecological indicators may be land use and land cover changes, biomass quality and quantity, water quality and quantity, and soil fertility. Energy management becomes the key measure for achieving the best in terms of the desired objectives.

Land Use and Cover Change

By sequential monitoring of land use and land cover changes, using a variety of technologies such as remote sensing, aerial photography and ground truth measurement, one can get a reasonable estimate of land quality. Any conclusions that suggest land degradation in the long run, obviously are not sustainable. Thus, for land use decision, on an ecological plane, a plot level analysis, leading through ecosystem level, and eventually getting to the land-scale level analysis, becomes most obvious; a similar set of units, at social scale, would be a family, a village and a cluster of villagers in a given ecological landscape. This would ensure that the end-product is sustainable livelihood as a short-term strategy, leading to sustainable development of the region as part of a long-term strategy, through a participatory and interactive analytical framework.

Biomass Quality and Quantity

The quality of biomass generated on the land (biodiversity), and the quantity of biomass generated through rehabilitation measures of a given ecosystem, are important indicators. We need quality and quantity of biomass for multiple use, for ecosystem integrity as well as for sustainable use by humans.

Water Quality and Quantity

Any land use and cover change, and indeed any land-based human activity, if unsustainable, may get reflected in groundwater recharge and the water quality. Monitoring this through sequential recharge characteristics of the system and/or monitoring water quality in rivers, ponds and lakes in the area immediately would give an idea about the sustainability of a given developmental pathway chosen for that landscape.

Soil Fertility

Sustainable management of soil fertility implies the effectiveness with which one is able to maintain soil fertility by strengthening internal biological processes within an ecosystem, both within the soil profile and above ground, and thus minimizes energy subsidies through inorganic fertilizers. The lesser the energy subsidy, the more sustainable the soil fertility management becomes.

Economic Indicators

Economic indicators are very much standardized in the already prevalent developmental paradigm. Monetary output/input analysis, capital savings of asset accumulation and dependency ratio are some of the measure adopted for economic evaluation. On a macroscale, the gross national product (GNP) is a unified indicator of national development level.

Social Indicators

In a social sense, 'quality of life' is a composite indicator, which is difficult to monitor, and even more difficult to reconcile with the subunits contained within. More easily measurable social indicators are health and hygiene, nutrition, food security, morbidity symptoms, educational status, etc. It is difficult to quantify indicators such as societal empowerment, and the less tangible ones in the area of socio-cultural values.

The best way to approach sustainable development is an integrated approach through full understanding of linkages between the ecological and social process.

To cope with increased demands of the basic requirement of life and the limited supply of the natural resources, along with consideration of environmental degradation and ecological balance, we need to make adjustment and adaptations with these complex global phenomena, and move towards sustainability of the biosphere. These were the considerations that moved an international group of ecologists from 14 countries to come together in Cuernavace, Mexico and propose an interdisciplinary cooperative research programme – the International Sustainable Biosphere Initiative – in the early 1990s. These were considerations that prompted the US National Research Council to set an agenda for action, which will enable the humans to complete the transitional journey towards sustainability. The council has developed an 'Integrated and Location-Specific Approach for Protecting Sustainability' in the following form:

Reporting on the Transition

Indicators are essential to inform society about how, and to what extent, progress is being made in navigating a transition towards sustainability.

It needs to be understood that multiple indicators are needed to chart progress towards the goals for meeting human needs and preserving life support systems, and to evaluate the efficacy of actions taken to attain these goals.

Integrated Knowledge and Action

Since the pathway to sustainability cannot be charted in advance, it will have to be navigated through trial and error and conscious experimentation. The urgent need is to design strategies and institutions that can better integrate incomplete knowledge with experimental action into programmes of adaptive management and social learning.

Priorities for Research Sustainability Science

Develop a research framework that integrates global and local perspectives to shape a place-based understanding of the interactions between environment and society.

Initiate focused research programmes on a small set of questions under study that are central to a deeper understanding of interactions between society and the environment.

Promote better utilization of existing tools and process for linking knowledge to action in pursuit of a transition to sustainability.

Priorities for Action: Knowledge-Action Collaborative

- Accelerate current trends in fertility reduction.
- Accommodate an expected doubling or tripling of the urban system in habitable, efficient and eco-friendly manner.
- Reverse declining trends in agricultural production in Africa.
- Accelerate improvements in the use of energy and materials.
- Restore degraded ecosystem while conserving biodiversity elsewhere.

Towards Sustainability

Sustainable development is the need of the present time not only for the survival of mankind but also for its future protection. Unlike the other great revolutions in human history – the Green Revolution and the Industrial Revolution – the 'sustainable revolution' will have to take place rapidly, consciously and on many different levels and in many different spheres, simultaneously. On the technical level, for example, it will involve the sustainable technologies based upon the use of non-renewable, fossil fuels for technologies that take advantage of renewable energies like the sun, wind and biomass, the adoption of conservation and recycling practices on a wider scale, and the transfer of cleaner and more energy-efficient technologies to countries in the developing world.

On the political and economic levels, it will involve, among other things, the overhauling of development and trade practices which tend to destroy the environment, and the improvement of indigenous peoples, a fairer distribution of wealth and resources within and between nations, the charging of true cost for products which exploit or pollute the environment, and the encouragement of sustainable practices through fiscal and legal controls and incentives. On the social plane, it will involve a renewed thrust towards emphasis on the education and social liberation of women. On the environmental level, we are talking about massive afforestation projects, renewed research into an assistance for organic farming practices and biopest control, and the vigorous protection of biodiversity. On the informational level, the need is for data that will allow the development of accurate social and environmental accountancy systems.

The aim of ecologically sustainable development is to maximize human well-being or quality of life without jeopardizing the life support system. The measures for sustainable development may be different in developed and developing countries according to their level of technological and economic development. But developing countries, like India, can focus attention on the following measures:

- ensure clean and hygienic living and working conditions for the people;
- sponsor research on environmental issues pertaining to the region;
- ensure safety against known and proven industrial hazards;
- find economical methods for salvaging hazardous industrial wastes;
- encourage afforestation;
- find out substitutes for proven hazardous materials based on local resources and needs instead of blindly depending on advanced nations to find solutions;
- ensuring environmental education as a part of school and college curriculum;
- encourage use of non-conventional sources of energy, specially solar energy;
- as far as possible, production of environment-friendly products should be encouraged;
- use of organic fertilizers and other biotechniques should be popularized;
- environmental management is the key of sustainable development, and it should include monitoring and accountability; and
- need for socialization and also humanization of all environmental issues.

The prime need for sustainable development is the conservation; and the development policy should follow the given norms:

(i) Make all attempts not to impair the natural regenerative capacity of renewable resources and simultaneously avoid excessive pollution hampering the biospherical capacity of waste assimilation and life support system.

(ii) All technological changes and planning strategy processes, as far as physically possible, must attempt a switch from non-renewable to renewable resources use.

(iii) Formulate a phase-out policy for the use of non-renewable resources in general.

Thus, for a worldwide sustainable growth, there is need for efficient and effective management of available resources. In this field, the production of 'environment-friendly products' (EFP) is a positive step. With the industrialization and technological development, markets are flooded with products of daily consumption. They could, however, be a source of danger to health and damage to our environment. There is, thus, need to distinguish the more environmentally harmful consumer products from those which are less harmful, or have a more benign impact on the environment, right from the stage of manufacture through packaging, distribution, use, disposal and reusability or recycling.

Throughout the world, emphasis is now being put on the production of EFP. In India, plans are afoot to market EFPs with combined efforts of Bureau of Indian Standards, Ministry of Environment and Forest and Central Pollution Control Board. Since 1990, a scheme of labelling 'Eco Mark' has also been started. In its first phase, the items included in this were soaps, plastics, papers, cosmetics, colours, lubricating oil, pesticides, drugs, and various edible items. The objectives of the scheme are: (i) to provide an incentive for manufacturers and to reduce adverse environmental impact of their products, (ii) to reward genuine initiatives by products, (iii) to assist consumers to become responsible in their daily lives by providing them with information to take account of environmental factors in their purchase decisions, (iv) to encourage citizens to purchase products which have less harmful environmental impact, and (v) to improve the quality of the environment and to encourage the sustainable management of resources.

Not only in consumer goods production but in the field of energy production also, environment-friendly techniques of power generation can be used. For example, in power production from coal, Pressurized Fluidized Bed Combined Cycle (PFBC) technique is useful in which coal is burnt efficiently and cleanly in combined cycle plants.

To cope with increased demand of the basic requirement of life and the limited supply of the natural resources, along with consideration of environmental and ecological balance, we need to emphasize on optimal management of land, water, minerals and other natural resources. There is also a need to utilize the native wisdom of those people, who live close to nature and earth, for eco-restoration along with development.

In order to apply the principle of sustainable management in reality, a highly complex way of looking at the problem is required, involving various disciplines. Sustainable is first and foremost a mental question. Without a grasp of the need or the will to change awareness, we will not succeed in realizing the principle of sustainability in agriculture. It is upon the decision-makers in politics to create the right framework and precondition for a sustainable development in agriculture. Global involvement, on the other hand, must not be left out of account. Sustainability reflects our understanding of necessity and responsibility about the question for whom, for what and how production can be guided into the future in a way that is efficient, environmentally sound and sparing on resources.

Global change is an ecological phenomenon, whereas globalization is concerned with economic change. A recent analysis of sustainable agriculture in the context of trade liberalization and globalization raises equally significant concern for a more informed decision-making process at local, regional and international levels. The emerging issues related to the impact of globalization on sustainable agriculture are as follows:

- There are explicit problems with the conventional theoretical economic condition for agriculture sustainability, especially when applied at the global level.
- The processes of trade liberalization and globalization will not be uniform given the ecological and institutional diversity of the nations of the world.
- There will be disparities in globalized impacts between rich and poor countries – for agriculture, industries, sustainability and environment as well as income and poverty.
- There is need for serious analysis of problems and policy initiatives, since the risk of disruption, social disruption and dislocation in the poorer countries of the world is clearly very high.

- The type of production technology research, facilitated by private research, will not address the significant public good and externality issues facing developing countries.
- There is need to focus on local farming situation as a basis of dealing with global problems, especially in poor countries.
- There is need to understand local institutional situations so as to determine appropriate remedial economic policies based on institutional sustainability.
- Integrated approach is essential for research and action at the regional scale related to water, atmosphere and climate, and species and ecosystem.

The pursuit of sustainability demands choices about the distribution of costs and benefits in space and time. There is also a need to take advantage of the 'traditional ecological knowledge' (TEK), which encompasses all issues related to ecology and natural resource management, both at local and regional levels. Along with political dimensions of environment-society relations, the TEK can be used for both eco-restoration and sustainable development.

Economics of Sustainable Development

The development and the environment have a two-way relationship: economic growth can have both good and bad impacts on environment. Environmental economics has sought to extend neo-classical theory, by encompassing the environment and attaching monetary values to losses in natural capital (Radclift,1993). The mainstream of sustainable development is based on a set of ideas that Low and Gleeson (1988) label as 'market environmentalism'.

Environmental economics like that of Pearce (1985) has advanced thinking of economics in a number of important ways, such as:

(a) They recognize that changes in natural capital stocks involve both costs and benefits.

(b) There is also recognition that both 'non-use' and 'use' values are important. Environmental values include 'existence values' which can be arrived at nationally, with a (theoretical) price attached to them.

(c) Environmental economics also recognizes uncertainty and irreversibility as principles in resource conservation.

(d) It is also recognized that 'optimality' refers only to economic uses and efficiency, and effectively excludes social goals for conserving resources.

Pearce (1989: 8) further declares that 'we know natural capital is valuable because people are willing to pay to preserve it ...'. He also observed that, 'Whether there is an actual market in the asset or not is not of great relevance. We can still find out what people would pay if only there were a market'.

Using economic concept, sustainability can be defined as constant stocks of both man-made and natural capital maintained over time. Economic development has always implied increasing capital stocks, but insights of environmental economics suggests the need to account for stocks of both man-made and natural separately. An alternative approach allowed the possibility of trade-offs between losses to natural capital in one project and gains elsewhere.

Eco-Auditing or Environmental Audit

Eco-auditing or environmental audit is a process of extracting information about a company that provides a realistic assessment, and thus helps in sustainable development. Eco-auditing is a systematic multidisciplinary method used periodically to assess the environmental performance of a project. Eco-auditing evolved as a management tool in the USA in 1980s. It has been promoted in Europe by International Chamber of Commerce and some multinational corporations as a means of getting effective environmental management for sustainable development.

The eco-audit includes the following:

(i) internal audit;
(ii) external audit;
(iii) aspects and impacts analysis;
(iv) applicable legislations;
(v) environmental objectives and targets;
(vi) environmental policy statement; and
(vii) corporate environmental plan.

Financing Sustainable Development

The world is facing serious environmental problems, which can be solved and controlled only when proper finances are available. The question is – who will pay? Naturally, those who are responsible for it, but at the same time, it is also the responsibility of those developed countries, which are capable both technologically as well as financially. Efforts have been made in this direction but still there is a long way to go.

Accepting sustainable management as an objective has been easier to achieve than the contentious issues of financing it. The available financial estimates vary in quality, and should be treated only as indicative of order of magnitude estimates. They do, however, provide the starting point for analysing financial requirements for sustainable estimates. This lack of robustness of the estimates also makes firm international agreements difficult to achieve.

Financing sustainable development involves the following three interrelated issues:

- the level or total amount of funding needed, depending on the scope of the overall programmes and targets to be achieved in specific areas;
- terms and conditions involved in the sharing of costs between the developed (high income) and the developing (low and middle income) countries; and
- countrywise mobilization of the requisite internal resources.

Funding necessary to provide sustainable development in the developing countries has been estimated in three different ways as discussed below:

Estimate 1

The two conventions of UNCED (Framework Convention of Climate Change and Convention on Biological Diversity) established the principle that the developing countries should be refunded the full agreed incremental costs associated with measures taken under them with Global Environmental Facility (GEF) as the interim funding mechanism. However, as environmental protection and development costs have proved to be difficult to define and operationalize, so the two conventions did not mention any specific cost.

The GEF was endowed with an initial funding of $1.3 billion for a three-year period from July 1991 to June 1994. It concentrates on four key areas: global warming, biological diversity, pollution of international waters, and stratospheric ozone depletion. Since March 1994, the GEF has become a permanent financial mechanism.

Estimate 2

The UNCED secretariat estimated the average annual (over the 1993–2000 period) cost of implementing Agenda 21 programmes for the developing countries. The estimates were based on incomplete supporting documents and not reviewed by the national governments. The estimated annual cost was just below $600 billion, with $130 billion (22%) provided through grant or concessional requirements. This amount is put in perspective when the GDP of all low-income and middle-income countries is noted; $5,276 billion in 1994. The $470 billion ($600–$130) cost to be borne by the developing countries is 8.9 per cent of their 1994 GDP.

Estimate 3

The 1992 World Development Report was devoted to the developmental and environmental efforts. It estimated that in 2000 AD, the additional investment in selected environmental programmes would amount to $75 billion or about 1.4 per cent of the GDP of developing countries for that year. The major items of additional investments are wastes from industry ($10–15 billion) and family planning ($7 billion). The report also recommended that such investments should be made on a self-financing basis to the extent consistent with equity and other objectives.

The above three examples of financial estimates suggest that clarity in this area is much below the desired level, in spite of its crucial importance. The methods used to arrive at these estimates not always being explicit, their robustness cannot be determined.

Financial resources and mechanism play a key role in the implementation of Agenda 21. In general, the financing for the implementation of Agenda 21 will come from a country's own public and private sectors. For developing countries, Official Development Assistance (ODA) is the additional funding for sustainable development. Renewed efforts are essential to ensure that all sources of funding contribute to economic growth, social development and environmental protection.

Box 4.4
Principles of Agenda 21

Human beings are at the centre of concerns for sustainable development. They are entitled to a healthy and productive life in harmony with nature.

States, in accordance with Charter of the United Nations should pursue their own environmental and development policies and make sure that they do not damage the environment of other states or of areas beyond the limits of national jurisdiction.

The right to development must be fulfilled so as to equitably meet development and environmental needs of present and future generations.

In order to achieve sustainable development, environmental protection shall constitute an integral part of the development process and cannot be considered in isolation from it.

All states and all people shall cooperate in the essential task of eradicating poverty as an indispensable requirement for sustainable development.

The special situation and needs of developing countries, particularly the least developed and those most environmentally vulnerable, shall be given special priority.

States shall cooperate in a spirit of global partnership to conserve, protect and restore the health and integrity of the earth's ecosystem.

To achieve sustainable development and a higher quality of life for all people, states should reduce and eliminate unsustainable patterns of production and consumption and promote appropriate demographic policies.

States should cooperate to strengthen endogenous capacity, i.e. building for sustainable development by improving scientific understanding through exchanges of scientific and technological knowledge.

Environmental issues are best handled with the participation of all concerned citizens at the relevant level.

States shall enact effective environmental legislation.

States should cooperate to promote a supportive and open international economic system that would lead to economic growth and sustainable development in all countries.

States shall develop national law regarding liability and compensation for the victims of pollution and other environmental damages.

States should effectively cooperate to discourage or prevent the relocation and transfer of any activities and substances to other states that cause severe environmental degradation or are found to be harmful to human health.

In order to protect the environment, the precautionary approach should be widely applied by states according to their capabilities.

National authorities should endeavour to promote the internalization of environmental costs and the use of economic instruments, taking into account the approach that the polluter should in principle, bear the cost of pollution.

Environmental impact assessment, as a national instrument, shall be undertaken for proposed activities that are likely to have a significant adverse impact on the environment.

States shall immediately notify other states of any natural disasters or other emergencies that are likely to produce sudden harmful effects on the environment of those states.

States shall provide prior and timely notification and relevant information to potentially affected states on activities that may have a significant adverse transboundary environmental effects.

Women have a vital role in environmental management and development. Their full participation is therefore essential to achieve sustainable development and ensure a better future for all.

The creativity, ideals and courage of the youth of the world should be mobilized to forge a global partnership in order to achieve sustainable development and ensure a better future for all.

Indigenous people and their communities and other local communities have a vital role in environmental management and development because of their native knowledge and traditional practices. States should recognize and duly support their identity, culture and interests and enable their effective participation in the achievement of sustainable development.

The environment and natural resources of people under oppression, domination and occupation shall be protected.

Warfare is inherently destructive to sustainable development. States shall, therefore, respect internal law providing protection for

> the environment in times of armed conflicts and cooperate in its further development, as necessary.
>
> Peace, development and environmental protection are interdependent and indivisible.
>
> States shall resolve all their environmental disputes peacefully and by appropriate means in accordance with the Charter of the United Nations.
>
> States and people shall cooperate in good faith and in a spirit of partnership for the fulfilment of the principles embodied in this Declaration and for the future development of international law in the field of sustainable development.

Agenda 21: Programme of Action for Sustainable Development

Agenda 21 is a programme of action for sustainable development worldwide. This is also known as the Rio Declaration on Environment and Development, adopted by more than 178 governments at the United Nations Conference on Environment and Development, known as the Earth Summit, held in Rio de Janerio, Brazil, from 3 to 14 June 1992.

Agenda 21 is intended to set out an international programme of action for achieving sustainable development in the 21st century. It stands as a comprehensive blueprint for action to be taken globally by governments, United Nations Organization, development agencies, non-governmental organization and independent sector groups, in every area in which human activity impacts the environment. It provides a broad review of issues pertaining to sustainable development, including statements on the basis of action, objectives, recommended activities and means of implementation.

Structure

Agenda 21 is a report of over 500 pages, comprising 40 chapters, organized into four sections, which addresses the major areas of political actions as follows:

Section I: Social and Economic Development

1. International cooperation to accelerate sustainable development in developing countries and related domestic policies

2. Combating poverty
3. Changing consumption pattern
4. Demographic dynamics and sustainability
5. Protecting and promoting human health
6. Promoting sustainable human settlement development
7. Integrating environment and development in decision-making

Section II: Conservation and Management of Resources for Development

1. Protection of the atmosphere
2. Integrated approach to the planning and management of land resources
3. Combating deforestation
4. Managing fragile ecosystem: combating desertification and droughts
5. Promoting sustainable agriculture and rural development
6. Conservation of biological diversity
7. Environmentally sound management of biotechnology
8. Protection of oceans and seas
9. Protection of the quality and supply of fresh water resources
10. Environmentally sound management of toxic chemicals
11. Environmentally sound management of hazardous wastes
12. Environmentally sound management of solid wastes and sewage-related issues
13. Safe and environmentally sound management of radioactive wastes

Section III: Strengthening the Role of Major Groups

1. Preamble
2. Global action for women towards sustainable and equitable development
3. Children and youth in sustainable development
4. Recognizing and strengthening the role of indigenous people and their communities
5. Strengthening the role of non-governmental organizations
6. Local authorities initiatives in support of Agenda 21
7. Strengthening the role of workers and their trade unions
8. Strengthening the role of business and industry

9. Scientific and technological community
10. Strengthening the role of farmers

Section IV: Measure of Implementation

1. Financial resources and mechanisms
2. Transfer of environmentally sound technology, cooperation and capacity building
3. Science for sustainable development
4. Promoting education, public awareness and training
5. National mechanism and international corporation for capacity building countries
6. International institution arrangements
7. International legal instrument and mechanism
8. Information for decision-making

The programme areas are generally structured in four heads as under:

- The basis for action
- The objectives
- The activities
- The means of implementation

In fact, Agenda 21 is a guideline for sustainable development. A programme for the further implementation of Agenda 21 was also adopted at the ninetieth special session of the United Nations General Assembly (Earth Summit +5), held from 23 to 28 June 1997 in New York by over 165 countries. This programme was designed to reconfirm the political commitment to sustainable development from all members of the international community, as well as from all major groups of civil society.

Millennium Development Goals (MDGs)

Another major step in sustainable development which covers the whole range of human activity and social development is the 'millennium development goals' set by the UN in the year 2000 for achievement by 2015. These are:

1. *Eradicate extreme poverty and hunger:* Reduce by half the 1.3 billion people living on less than $1 a day and the 815 million without enough food to meet daily energy needs.

2. *Achieve universal primary education:* 115 million children, mostly in sub-Saharan Africa and Southern Asia, do not go to school. Reduce this number.
3. *Promote gender equality and empower women:* In three main areas – education, employment and political decision-making.
4. *Reduce child mortality:* 11 million children die before the age of 5, most from treatable diseases. The aim is to cut this by two-thirds.
5. *Improve maternal health:* Particularly in sub-Saharan Africa, by investing in health facilities, birth attendants and education.
6. *Combat HIV/AIDS, malaria and other diseases:* Around 39.4 million people are living with HIV, and malaria kills more than 1 million people per year. The aim is to reduce these numbers significantly.
7. *Ensure environmental sustainability:* Reverse the loss of environmental resources – forests, biological diversity, the ozone layer and provide safe water, adequate sanitation and affordable housing.
8. *Develop a global partnership for development:* Encourage the North to increase aid to developing countries, to move to fairer trade and cancel the debts of struggling developing countries.

Environmental Issues and Sustainable Development

Man, with his technological and scientific skills, has made rapid developments in agriculture, mining, industries, transportation, forestry, land management and other areas. Unfortunately, in doing so, there has been much disruption of the functioning of natural environment. Now, people have begun to realize the side-effects of the developmental activities and have started their efforts in minimizing these effects by adopting methods and technology for sustainable development. Some of the important environmental issues associated with development, and the role of ecology in possible solution of these problems, i.e. sustainable development, are as follows:

Food Production and Organic Farming

The increasing demand of food for growing population has become a serious problem in many parts of the world, especially in developing countries like India, China, Indonesia, Egypt, Brazil, and they have been successful in increasing the quantity and quality of food and other crops. This has been possible due to application of chemical fertilizers, high yielding seeds, crop rotation practices, weed control, pesticides as well as due to mechanization in agriculture. This resulted in the Green Revolution beginning in the late 1960s with an attempt to increase food production in developing countries. New varieties of wheat, rice and corn were developed with better irrigation and use of chemicals for pest and weed control.

But along with the increase in food production, there came an increased dependence on energy-based expensive technology and several ecological problems have emerged due to excessive use of chemicals either in the form of fertilizers or as pesticides. Now, it has been realized throughout the world that use of chemical fertilizers and other chemicals is harmful to soil productivity and has not only disturbed the ecosystem but is also a cause of social revolution. In countries like USA, farmers are now shifting from routine use of herbicides, pesticides, fertilizers, etc. to organic farming in which no chemicals are used. The only fertilizers used are organic matters, such as manure, crushed limestone, etc. This system is similar to that of ancient Indian system which is more eco-friendly. The adoption of organic farming is a way for sustainable development. By adopting proper management practices, one can develop high yielding varieties even in this system. Along with agriculture, aquaculture can also be developed. The best way is to develop fish farming, which fulfils the increasing demands of foods, especially that of protein.

By adopting certain techniques, agriculture can become sustainable. Some of the agricultural technologies with high potential sustainability are as follows:

Intercropping – The growing of two or more crops simultaneously on the same piece of land. Benefits arise because crops exploit different resources, or interact with one another. This interaction may also serve to control pests and weeds.

Rotations – The growing of two or more crops on the same piece of land. Benefits are similar to those arising from intercropping.

Agro-forestry – A farming of intercropping in which annual herbaceous crops are grown with perennial trees or shrubs.

Sylvo-pasture – Similar to agro-forestry, but combines trees with grassland and other fodder species on which livestock graze.

Green manuring – The growing of legumes and other plants in order to fix nitrogen and then incorporating them in the soil for the following crops. Commonly used green manures are Sesbania and the fern Azolla, which contain nitrogen-fixing, blue green algae.

Conservation tillage – Systems of minimum tillage or no-tillage, in which the seed is placed directly in the soil with little or no preparatory cultivation.

Biological control – The use of natural enemies, parasites or predators, to control pests.

Integrated pest management – The use of appropriate techniques of controlling pests in an integrated manner that enhances rather than destroys natural controls.

Energy and Sustainable Development

Sustainable energy development (SED) involves the harnessing of energy resources for human use in a manner that makes development more sustainable. Energy production and use have strong links with the economic, social and environmental dimensions of sustainable development.

First, economic growth depends on energy services such as heating, refrigeration, cooking, lighting, communications, motive power and electricity.

Second, energy is a basic social need that affects human well-being. Wide disparities in access to affordable commercial energy and energy services in both urban and rural areas threaten social stability. Access to decentralized small-scale energy technologies is an important element for successful poverty alleviation. Women and children, who are relatively more dependent on traditional fuels, suffer disproportionately.

Third, there are many environmental links (mainly negative), since energy production and use continues to be a primary source of local, transnational and global pollution. Specific impacts include: groundwater and air contamination; land degradation and changes in use; marine and costal pollution; ecosystem destruction and loss of biodiversity; damage to health, structures and natural systems from SO_2 and NOx and ash particulates which degrade air quality; and, finally greenhouse gas emissions which have harmed the global environment.

Energy for sustainable development will require: the more efficient use of energy, especially at the point of end use in buildings; transportation and production processes; increased reliance on renewable energy sources; and accelerated development and deployment of new energy technologies. Some strategies and policies for globally achieving SED include: encouraging greater international cooperation in areas such as technology procurement, harmonization of environmental taxes and emission trading, and energy efficiency standards for equipment and products; adopting policies and mechanism to increase access to energy service through decision; advancing innovation, with balanced emphasis on all steps of the innovation chain; encouraging competitiveness in energy markets to reduce total costs of energy services to the users; cost-based prices, including phasing out all forms of subsidies for fossil fuels and nuclear power and internalizing external environmental and health impacts; encouraging greater energy efficiency; and developing and/or diffusing new technologies to wider markets.

In the ecosystem, energy plays a very important role. The sun is the original source of energy on our planet. All forms of energy, i.e. nuclear, wind, coal, electric, water, gasoline, etc. originate from the sun. The total energy received by the earth's surface from the sun is enormous and much more than required by human beings. But, in modern times, the use of fossil fuels, hydroelectricity, and to some extent nuclear fission are the sources of energy.

The rapid exploitation of these energy sources has resulted in an energy crisis. In coming times, fossils fuels such as oil, gas and coal will become scarce. The impact of big dams and thermal as well as atomic power stations is a cause for concern to environmentalists. Therefore, for sustainable development, it is necessary to adopt non-conventional sources of energy like solar, wind, tidal, etc. Of these, the best is solar energy, provided that we are able to harness it with the cheapest possible technology. In tropical countries like India, solar energy is the only alternative, which will provide an energy without any adverse impact on environment. Steps have already been taken for the development of photovoltaic cells, solar-energized water pumps, air conditioning units, solar water heaters, solar home-heating units, solar cookers, etc. Research in this field is also going on. What is needed is the development of cheaper technology to harness the solar and other non-conventional sources of energy.

Energy is directly related to the most critical social issues which affect sustainable development. These are poverty, jobs, income levels, access to social services, gender disparity, population growth, agricultural production, climate change and environmental quality, and economic and security issues. Without adequate attention to the critical importance of energy and all of these aspects, the global, social, economical and environmental goals of sustainability cannot be achieved. Indeed, the magnitude of change needed is immense, fundamental and directly related to the energy produced and consumed internationally.

World primary energy demand is projected to expand by more than one-half between 2005 and 2030, or 1.6 per cent annually (IEA, 2005). Over two-thirds will come from DCs (developing countries) where per capita use is low. At present, DCs comprise over three-quarters of the world's population but utilize only one-quarter of the world's energy. Most of this population live in rural areas, have little or no access to commercial energy, and rely heavily on traditional biomass fuels such as wood, crop waste and animal dung.

Energy demand has been growing rapidly in DCs in recent decades, creating supply shortages. Making energy development more sustainable in the developing world requires reconciliation of economic growth (including increased use of energy) and poverty alleviation effort, with responsible stewardship of the environment. This has to be done without overburdening the already weak economics and without diverting funds from other key goals such as education and healthcare.

The use of non-conventional and renewable energy is necessary for sustainable development. These are as follows:

Small Hydro Projects

Mini- and micro-hydro plants are popular for providing local power supply and irrigation without the massive environmental and social impacts arising from large hydro schemes. Micro-hydro projects are currently underway in many DCs, which have the advantage of providing electricity to rural areas beyond the central grid.

Solar Energy

Advances in solar thermal technologies and direct electricity generation by photovoltaics (PVs) have reduced costs significantly since the late 1990s. Solar energy is constrained by its limited applicability.

Geothermal Energy

Geothermal energy uses the earth's heat to drive steam turbines. Cost-competitive exploitation of geothermal energy with current technology is largely limited to the volcanically active Pacific 'Ring of Fire' and the Mediterranean, where suitable steam reservoirs are located within one mile of the surface. Geothermal steam carries with it a number of atmospheric pollutants, including carbon dioxide, mercury and radon. Currently, the toxins are commonly re-injected into the reservoir. With recent advances, hot dry rock technology will be competitive with conventional geothermal technology and fossil-fuelled plants.

Biomass Power Generation

A wide range of options exist for converting biomass into electric power. Most potential applications are for remote power supply to DCs. Concerns about the effects of dedicated fuelwood plantations on the local environment and competing land uses impose constraints on the widespread adoption of dendro-thermal power supply schemes. It is economical to use biomass for energy purpose only where it is available as a byproduct of other processes. Another source of biomass for power generation is municipal solid waste, whose use is limited both by public concern over emissions and the shortage of concentrated sources of solid waste.

Wind Power

Large grid-connected wind generators (75 kw – 450 kw) in multi-turbine 'wind farms' now supplement power grids in many countries. The total installed wind-power capacity was about 47,300 MW in 2004, with Germany, Spain and the USA accounting for 3.5 per cent, 18 per cent and 14 per cent, respectively (MIT, 2005). However, wind energy provided only 0.3 per cent of the total electricity supply.

Ocean Energy

Pilot plant experience with ocean thermal energy conversion (OTEC) indicates that temperature differences of 20°C between surface waters and waters about 1000 m below are sufficient for economic power generation. There is a vast potential for this energy source in the tropical oceans. Due to high capital requirements and the unproven technology, no large commercial plants are functioning.

Problem of Waste Disposal and its Management

In developed countries, enormous use of commodities in daily life is causing environmental problems. This problem is now not only limited to developed countries but is growing day by day in developing countries also. Wastes and toxic substances are accumulating at a rapid rate in our surroundings. These substances enter the biogeochemical cycles. In man-made ecosystem, various types of wastes accumulate in the environment which upset the growth of microorganisms. A variety of synthetic matter like plastic, rubber, glass, etc. are not degradable, thus, causing environmental problems. Wastes include domestic wastes, industrial and mining wastes and agricultural wastes.

Various types of wastes, if properly managed, can be used and steps have already been taken in this direction. Through gene-engineering, new strains of bacteria and other microorganism are being bred, which can help to decompose various harmful industrial effluents into non-toxic agents.

Scientists have also developed methods and technology to utilize sewage waste for wood, cattle feed, fertilizer and fuel production. Municipal sewage treatment plants remove large objects, floating matter and settable solids and also through physico-chemical and biological processes break down the organic matter and remove harmful microbes before release of the treated sewage into water bodies.

Due to the lack of proper facilities, sewage is allowed to enter nearby water bodies without any treatment. The untreated sewage causes serious water pollution and also damages the local ecology. It is, therefore, high time that sewage is recycled in an economically feasible and environmentally safe manner to meet the various needs of mankind. Several techniques have been developed for sewage treatment and to extract water for irrigation, fodder, fuel and fertilizer from it. Efforts are also being made to use sewage on a large scale to improve farmlands, reclaim spoilt areas and create new recreational facilities. Use of sewage water for fish farming is common in countries like Germany, Poland, Russia, Hungary, Israel, China, USA, Indonesia, etc. Sewage is also being exploited for fuel, like the methane gas released due to bacterial growth can be used as fuel. There is thus an urgent need to develop an efficient waste management system.

Irrigation and Problem of Salinity and Waterlogging

During the last fifty years, vast areas have been provided with irrigation facilities. This has definitely increased the agricultural production throughout the world. But, due to the lack of water management and also due to illiteracy of the farmers and the geological structure of a particular region, vast areas are now suffering from the problem of salinity and waterlogging.

Due to the gradual depletion of organic matter and microbial population involved in decomposition, there is increase in salinity and waterlogging, which have adverse effects on soil fertility. Waterlogging and increased salinity are mostly the outcome of poor drainage and mismanagement of water in areas covered by large irrigation projects. The recent example is the Ganganagar and Bikaner districts of Rajasthan, where problem of waterlogging has become serious and several villages have been evacuated and there is no possibility of crop production in these areas. This has happened due to the irrigation water of the Indira Gandhi Canal. In Punjab, the seepage from unlined canals as well as due to irrigation has raised the water table by 7–10 metres above the previous levels. Similar is the problem of salinity, which has damaged the irrigated land. These problems can be checked by a proper water management system.

Deforestation and Desertification

The problem of deforestation has arised mainly due to destruction of forests for various developmental or commercial purposes. The growth of population, rapid expansion of cities and other settlements, demands for land for new agricultural farms, mining, construction of road, dams, etc. have all resulted in a mass cutting of natural vegetation. Many useful plants have been made extinct and several thousands are nearing extinction. In the absence of reforestation, the forest cover throughout the world is declining at a rapid pace. The dense tropical forests too are suffering from the problem of mass cutting for commercial purpose.

Box 4.5
India's Efforts towards Sustainable Development

I. In WSSD-2000, India presented its strategy for sustainable development, entitled, 'Empowering People for Sustainable Development' (EPSD). The objectives of EPSD were: combating poverty, empowering people, using core competence in science and technology, and setting environmental standards.

II. National Action Plan for Climate Change (NAPCC): In June, 2008 India launched NAPCC to combat climate change through a policy of sustainable development. The following are the main aspects of this action plan:
   - Protecting the poor through sustainable development strategy.
   - Achieving objectives of national growth and poverty alternatives through ecological sustainability.
   - Appropriate technology for adaptation and mitigation.
   - Effective implementation through unique linkages – with civil society, local governments and public-private partnerships.

Under this action plan, eight national missions for India's sustainable development have been set up. These are:

1. National Solar Mission
2. National Mission for Enhanced Energy Efficiency
3. National Mission on Sustainable Development
4. National Water Mission
5. National Mission for Sustaining the Himalayan Ecosystem
6. National Mission for a Green India
7. National Mission for Sustainable Agriculture
8. National Mission on Strategic Knowledge for Climate Change.

In order to protect the environment and ecosystem, action plans should be implemented not only to protect the present forest cover but also for large-scale afforestation, so that 30 to 33 per cent of the land area can be developed as natural and/or planted forests. There is also a need for the enforcement of present laws strictly to avoid illegal and indiscriminate felling of trees by businessmen, industrialists and rural masses. Both social and agro-forestry should be encouraged and wastelands be developed as forest lands. Besides these, public participation should also be encouraged at all levels.

The threat of desertification and expansion of desert area is also due to overgrazing, loss of natural vegetation and other human activities. UNESCO studies have shown that approximately 43 per cent of total land area in the world has been affected by desertification. The suggested measures for the check of desertification are improvement in land use patterns, stabilization of sand dunes, development of natural vegetation suitable to arid climate and reduction of population pressure, etc.

Sustainable development should essentially be based on maintaining the fragile balance between productivity functions and the ecosystem. For sustainable development, our approach should be holistic for every resource sector and at all levels of local, regional and global scale. Unless we adopt this approach for sustainable development, the problem of environmental degradation cannot be solved.

MULTIPLE-CHOICE QUESTIONS

1. In which report/declaration the term 'sustainable development' was first used?

   (a) Rio Declaration

   (b) Brundtland Report, 'Our Common Future'

   (c) Cocoyoc Declaration on Environment and Development

   (d) Stockholm Declaration on Human Environment

2. Sustainable development is described as the development that meets the needs of the present without compromising the ability of future generations to meet their own needs. In this perspective the concept of sustainable development is intertwined with which of the following concepts?

   (a) Globalization (b) Empowerment

   (c) Inclusive growth (d) Social justice and empowerment

3. Consider the following statements:

   1. The consumption of natural resources is increasing at an alarming rate, causing great concern for their conservations.
   2. Developmental activities have become a big threat to the environment.
   3. The degradation of environment has caused great concern, and a big question is – what will happen next?

Which of the above statement(s) has have given the basic ideas of sustainable development?

(a) Only 1 (b) Only 2

(c) 1 and 2 (d) 1 and 3

4. The concept of 'Sustainable Development' was elaborated first time in a report entitled:
   (a) Our Common Future
   (b) Agenda 21
   (c) Millennium Development Goals
   (d) Economics of Sustainable Development

5. Which of the following statements is not related with sustainable development?
   (a) An environmental 'handrail' to guide for economic development
   (b) Only traditional knowledge be used for economic development.
   (c) Balancing the fulfilment of human needs with the protection of the natural environment.
   (d) Improving the quality of human life while living within the carrying capacity of supporting ecosystems

6. Better quality of life in cities and communities can be developed through:
   (a) development of agriculture
   (b) social and cultural development
   (c) long-term environmental development
   (d) conservation of resources

7. The main component(s) of sustainable development matrix is/are:

   (a) Planet Earth (b) People

   (c) Profit (d) All the above

8. Which is not an element of profit in sustainable development matrix?

   (a) Energy flow (b) Create values
   (c) Eliminate waste (d) Recognize interdependence

9. Gifford Pinchot, the former US President, Roosevelt's chief forester and mentor advocates the strategy of 'sustainable utilization' of which aspect?

   (a) Conservation of biodiversity
   (b) Conservation of resources
   (c) Soil conservation
   (d) Conservation of water resources

10. Which of the following report have generated a debate on sustainable development in the world?

   (a) Resourceful Earth (19984)
   (b) Global Possible (1985)
   (c) Our Common Future (1987)
   (d) US Global 2000 Report (1980)

11. Who developed the concept of Zero growth?

   (a) Daly (1977) (b) Norgaard (1985)
   (c) Pinchot (1910) (d) Simon & Kohn (1984)

12. Norgaard has developed the approach of environmental economics. Which concepts was suggested by him in this approach?

   (a) Natural resource management
   (b) Environment-friendly techniques
   (c) Cost-benefit techniques
   (d) Organic agriculture

13. Which development is considered as Rio Declaration?

(a) UN Declaration (b) Global Report
(c) Agenda-21 (d) Development Report

14. The Millennium Ecosystem Assessment describes the following major categories of ecosystem services: provisioning, supporting, regulating, preserving and culture. Which one of the following is supporting service?

(a) Production of food and water
(b) Control of food and water
(c) Nutrient cycling and crop pollution
(d) Maintenance of diversity

15. Which of the following is not one of the International Development Targets of the Millennium Development Goals?

(a) Reducing the number of persons living in extreme poverty by one half by 2015.
(b) Universal primary education by 2015.
(c) Reduction of infant and child mortality by 2/3rd by 2015.
(d) Reducing the total population by 1/3rd

16. Which of the following is not a main issue of Agenda-21?

(a) Biodiversity (b) Sustainable development
(c) Employment (d) Water quality

17. What was the main concern of the Brundtland Report?

(a) Gross National Product
(b) Foreign investment
(c) Concern of Present generation about future
(d) International Coordination

18. What refers to 'Caring for the Earth' in Agenda-21?
    (a) Caring of women
    (b) Caring of children
    (c) Caring of indigenous people
    (d) All the above

19. Which programme was adopted at the close of Earth Summit?
    (a) Resource conservation strategy
    (b) Programme for the further implementation of Agenda-21
    (c) Sustainable development programme
    (d) All of the above

20. Which of the following targets have been set in 2002 Johannesburg Earth Summit?
    (a) Sanitation (b) Energy use
    (c) Health (d) All the above

21. Which is not a Millennium Development Goals (MDGs)?
    (a) Primary universal education
    (b) Reduce child mortality
    (c) The urban challenges
    (d) Improve material health

22. What is common in three documents in Brundtland report, Agenda-21 and MDGs?
    (a) Plan for conservation for resources
    (b) Close-knit relationship between international development and sustainable development
    (c) Promotion of gender equality and empower women
    (d) None of the above

23. Which statement(s) is/are true about eco-development?

1. The development based on regional ecological balance.
2. Every type of pollution is in control state.
3. Human resources are utilized to the maximum
4. Development at regional and local levels in accordance with ecosystem

Code:

(a) Only 1 (b) 1, 2 and 3
(c) Only 4 (d) 1 and 4

24. Dasman (1985) described eco-development as a triangle. Which is not a side of eco-development triangle?

(a) Resource conservation (b) Basic Needs
(c) Self-reliance (d) Ecological Sustainability

25. Which of the following is a sustainable activity?

(a) Intensive cultivation of land without adequate care of soil fertility
(b) Development of green belt around cities.
(c) Development of irrigation facilities without proper water management
(d) Replacement of high yielding varieties

26. Which is not a proper step towards sustainability?

(a) From non-renewable to renewable resource strategy
(b) A production efficiency era of minimum environmental damage
(c) Excessive use of non-degradable material
(d) A stable world population

27. Which of the following is not an ecological indicator of sustainability?
    (a) Soil fertility
    (b) Water quality and quantity
    (c) Biomass quality
    (d) Mineral resources

28. Which of the following document introduces the essential framework for sustainable development in India?
    (a) National Plan for Environmental Development
    (b) Empowering People for Sustainable Development
    (c) National Action Plan for Climate Change
    (d) National Mission on Sustainable Habitat

29. In which year did India formulate the National Action Plan for Climate Change?
    (a) 2002 (b) 2005
    (c) 2008 (d) 2010

30. Which one of the following is not a national mission for India's sustainable development?
    (a) National Water Mission
    (b) National Mission for Pollution Control
    (c) National Mission for a Green India
    (d) National Mission for Sustainable Agriculture

31. Where was United Nations Conference on Sustainable Development, 2012 held?
    (a) Nairobi (b) Geneva
    (c) Vienna (d) Rio de Janeiro

32. 'Green Economy' is mainly concerned with:

(a) Agriculture

(b) Forest development

(c) Human well-being and social equity

(d) Resource conservation

33. Rio+20 Declaration has adopted a resolution, which is known as?

(a) Green Economy (b) Save Earth

(c) Green Agenda (d) The Future We Want

34. The primary concern of 'Green Agenda' is:

(a) Concern for environment and for the future generations.

(b) Concern for the development for forest

(c) Concern for development of agriculture and irrigation

(d) Concern for global awareness about environment degradation

35. Consider the following statements:

1. Eradicating poverty is the primary requirement for sustainable development
2. Sustainable development is possible by providing employment to all
3. Sustainable development can be promoted by reducing inequalities

Which of the following statement(s) is are included in Rio+20 Declaration?

(a) Only 1 (b) Only 2

(c) 1 and 3 (d) 1, 2 and 3

36. Which is not true about ecologically sustainable development?
    (a) Maximize human well-being without jeopardizing the life support system
    (b) Encourage use of non-conventional sources of energy
    (c) Ensure clean and hygienic living and working conditions for the people
    (d) Encourage chemical fertilizers for agricultural production

37. Which scheme was introduced by Bureau of Indian Standards, Ministry of Environment and Forest and Central Pollution Control Board in 1990 for environmental friendly products?
    (a) Green Labelling (b) Ecomark
    (c) Red Labelling (d) All of these

38. Which of the following agriculture/farming system is more sustainable?
    (a) Organic farming (d) Aquaculture
    (c) Agro-forestry (d) All of these

39. Which is not a sustainable technique in agriculture.
    (a) Single cropping
    (b) Agro-forestry
    (c) Crop Rotations
    (d) Integrated Pest Management

40. Which is not a part of sustainable energy development?
    (a) More efficient use of energy
    (b) Use of hydro and atomic power
    (c) Increased reliance on renewable energy sources
    (d) Accelerated development of new energy technologies

41. National Green Tribunal Act, 2010 was enacted in consonance with which of following provisions of Constitution of India?

1. Right to healthy environment construed as a part of life under Act 21.
2. Provisions of grants for raising the level of administration in the scheduled areas for the welfare of scheduled tribes under Act 275c.
3. Powers and functions of Gram-Sabha as mentioned under Act 243.

Which of the statements given above is/are correct?

(a) Only 1 (b) 2 and 3

(c) 1 and 2 (d) All of these

Note: Question Number 42 to 45 are based on **Assertion (A)** and **Reason (R)** select your answers using the following codes:

(a) Both **A** and **R** are true and **R** is not the correct explanation of **A**

(b) Both **A** and **R** true but **R** is not the correct explanation of **A**

(c) **A** is true **R** is false.

(d) **A** is false **R** is true.

42. **Ascertain (A):** Sustainable development is the development that meets the needs of present without compromising the ability of future generations to meet their own needs.

Reason (R): The consumption of natural resources is increasing at an alarming rate, causing great concern for their conservation.

43. **Ascertain (A):** There is a need for natural understanding between developed and developing countries so that sustainable development can be possible.

Reason (R): The environmental degradation and associated problems in the Third World, are consequences of present development policies.

44. **Ascertain (A):** Development of irrigation facilities without proper water management leads to waterlogging, alkaline or saline soil.

Reason (R): Eco-development plan for a region should be based on social issues.

45. **Ascertain (A):** There is also need to take advantage of the 'traditional ecological knowledge' (TEK), which encompasses all issues related to ecology and natural resource management.

Reason (R): Along with political dimensions of environment-society relations, the TEK can be used for both eco-restoration and sustainable development.

Answers

1. c	2. b	3. d	4. a	5. b	6. c	7. d
8. a	9. b	10. c	11. a	12. b	13. c	14. c
15. d	16. c	17. c	18. d	19. b	20. d	21. c
22. b	23. d	24. a	25. b	26. c	27. d	28. b
29. c	30. b	31. d	32. c	33. d	34. a	35. c
36. d	37. b	38. d	39. a	40. b	41. a	42. a
43. b	44. c	45. b				

Glossary

acid rain A condition in which natural precipitation becomes acidic after reacting chemically with pollutants in the air.

aerosol A substance consisting of small particles, typically having diameters that range from 1/100 micrometer to 1 micrometer.

air pollution The deterioration of the quality of air that results from the addition of impurities.

albedo A measure of the reflectivity of a surface, measured as the ratio of light reflected to light received. A mirror or bright snowy surface has a high albedo, whereas a rough, flat road surface has a low albedo.

arithmetic growth In population studies, growth characterised by the addition of a constant number of individuals during a unit interval of time. For instance, if there are x individuals in year O and x + a in year 1, arithmetic growth implies x + na in year n.

atmosphere The predominantly gaseous envelope that surrounds the earth.

atom The fundamental unit of an element.

atomic nucleus The small positive central portion of the atom that contains its protons and neutrons.

atomic number The number of protons in an atomic nucleus.

autotroph An organism that obtains its energy from the sun, as opposed to a heterotroph, which is an organism that obtains its energy from the tissue of other organisms. Most plants are autotrophs.

background radiation The level of radiation on earth from natural sources.

benthic organism A plant or animal that lives at or near the bottom of a body of water.

biodegradable Refers to substances that can readily be decomposed by living organisms.

biodiversity A term describing the variety of species, both flora and/or fauna, contained within an ecosystem.

biogas Methane produced by fermentation of organic waste. Biogas is chemically identical to methane produced in any other manner.

biomass The total weight of all the living organisms in a given system.

biome A group of ecosystems characterised by similar vegetation and climate that are collectively recognisable as a single large community unit. Examples include the Arctic Tundra, the North American prairie, and the tropical rain forest.

biosphere That part of the earth and its atmosphere that can support life.

birth cohort A group of individuals born in a given period, such as in a particular year.

birth rate The number of individuals born during some period, usually a year, divided by an appropriate population. For example, the crude birth rate in human populations is the number of live children born during a given year divided by the mid-year population of that year.

black lung disease A series of debilitating and often fatal diseases that affect the lungs of miners who work in underground coal mines.

bloom A rapid and often unpredictable growth of a single species in an ecosystem.

branching chain reaction A chain reaction in which each step produces more than one succeeding step.

breeder reactor A nuclear reactor that produces more fissionable material than it consumes.

cancer An abnormal, unorganised, and unregulated growth of cells in an organism.

carnivore An animal that eats the flesh of other animals.

carrying capacity The maximum number of individuals of a given species that can be supported by a particular environment.

chain reaction A reaction that proceeds in a series of steps, each step being made possible by the preceding one. See also branching chain reaction.

chemical bond A linkage that holds atoms together to form molecules.

chemical energy The energy that is absorbed when chemical bonds are broken or released when they are formed.

clear-cutting The practice of cutting all the trees in a designated area. Clear-cutting leaves behind barren, open regions but is sometimes the cheapest way to harvest timber if only short-term economic evaluation is used. See selective cutting.

climate The composite pattern of weather conditions that can be expected in a given region. Climate refers to yearly cycles of temperature, wind, rainfall, and so on and not to daily variations.

climax community A natural system that represents the end, or apex, of an ecological succession.

common mode failure An accident in which a single event causes multiple failures and thereby knocks out redundant systems. See redundancy.

competition An interaction in which two or more organisms try to gain control of a limited resource.

composting The controlled, accelerated biodegradation of moist organic matter to form a humus-like product that can be used as a fertiliser or soil conditioner.

cooling pond A lake or pond used to cool water from an electric generating station or any other industrial facility.

cooling tower A large tower-like structure used to cool water from an electric generating station or any other industrial facility.

critical condition In nuclear science, a critical condition is said to exist when a chain reaction continues at a steady rate, neither accelerating nor slowing down.

critical minimum size In general, a critical condition relates to a point at which some property changes very abruptly in response to a small change in some other property of the system. In ecology, a population is said to be reduced to its critical minimum size when its numbers are so few that it is in acute danger of extinction.

crude rate A vital rate with the entire population of some area as the denominator.

death rate The number of individuals dying during some period, usually a year, divided by an appropriate populations. For example, the crude death rate in human populations is the number of deaths during a given year divided by the mid-year population of that year.

decibel (*dB*) A unit of sound intensity equal to 1/10 of a bel. The decibel scale is a logarithmic scale used in measuring sound intensities relative to the intensity of the faintest audible sound.

demographic transition The pattern of change in vital rates typical of a developing society. The process can be outlined briefly as follows. Birth and death rates in a pre-industrial society are typically very high; consequently, population growth is very slow. Introduction or development of modern medicine causes a decline in death rates and hence a rapid increase in population growth. Finally, birth rates fall, and the population grows slowly once more.

demography The branch of sociology or anthropology that deals with the statistical characteristics of human populations, with reference to total size, density, number of deaths, births, migrations, marriages, prevalence of disease, and so forth.

desert A climate system in which rainfall is less than 25 centimeters per year. These systems are barren and support relatively little plant or animal life.

desertification The diminution or destruction of the biological potential of the land, which can ultimately lead to desert-like conditions.

doubling time The time a population takes to double in size, or the time it would take to double if its annual growth rate were to remain constant.

ecological niche The description of the unique functions and habitats of an organism in an ecosystem.

ecology The study of the interrelationships among plants and animals and the interaction between living organisms and their physical environment.

economic externality That portion of the cost of a product that is not accounted for by the manufacturer but is borne by some other sector of society.

ecology the science of all the relations of all organism and their environment.

ecosystem A group of plants and animals occurring together plus that part of the physical environment with which they interact. An ecosystem is defined to be nearly self-contained, so that the matter that flows into and out of it is small compared with the quantities that are internally recycled in a continuous exchange of the essentials of life.

ecosystem homeostasis The control mechanisms within an ecosystem that act to maintain constancy by opposing externals.

entropy A thermodynamic measure of disorder. It has been observed that the entropy of an undisturbed system always increases during any spontaneous process; that is, the degree of disorder always increases.

environment The sum total of all conditions, agencies and influences which affect the development, growth, life and death of an organism, species or race.

environmental resistance The sum of various pressures, such as prediction, competition, adverse weather, and so on, that collectively inhibit the potential growth of every species.

epidemiology The study of the distribution and determination of health and its disorders.

estuary A partially enclosed shallow body of water with access to the open sea and usually a supply of fresh water from the land. Estuaries are less salty than the open ocean but are affected by tides and, to a lesser extent, by wave action of the sea.

euphotic zone The surface volume of water in the ocean or a deep lake that receives sufficient light to support photosynthesis.

eutrophication The enrichment of a body of water with nutrients, with the consequent deterioration of its quality for human purposes.

expectation of life The number of years an infant can be expected to live under a specified schedule of death rates.

extrapolation The prediction of points on a graphout side the range of observation.

evolution, theory of A theory that states that species are not unchangeable but arise by descent and modification from pre-existing species.

fire climax A condition in which the continuance of a given ecosystem is maintained by fire.

fission (*of atomic nuclei*) The splitting of atomic nuclei into approximately equal fragments.

food chain An idealised pattern of flow of energy in a natural ecosystem. In the classic food chain, plants are eaten only by primary consumers, primary consumers are eaten only by secondary consumers, secondary consumers only by tertiary consumers, and so forth.

food web The actual pattern of food consumption in a natural ecosystem. A given organism may obtain nourishment from many different trophic levels and thus give rise to a complex, interwoven series of energy transfers.

fusion (*of atomic nuclei*) The combination of nuclei of light elements (particularly hydrogen) to form heavier nuclei.

Gaia The ancient Greek goddess of the earth. This word has recently been used to describe the biosphere and to emphasise the interdependence of the earth's ecosystems by likening the entire biosphere to a single living organism.

gas A state of matter that consists of molecules that are moving independently of each other in random patterns.

gasohol A motor fuel consisting of approximately 90 per cent gasoline and 10 per cent ethyl alcohol.

genetic engineering The process of artificially removing specific genes from one organism and replacing them with genetic information from another.

geo-metric growth In population studies, growth such that in each unit of time, the population increases by a constant proportion. Also called exponential growth.

geothermal energy Energy derived from the heat of the earth's interior.

global warming is the increase in global temperatures brought about by increased emission of 'green house' gases into the atmosphere.

green revolution The realisation of increased crop yields in many areas due to the development of new high-yielding strains of wheat, rice, and other gains in the 1960s.

green-house effect The effect produced by certain gases, such as carbon dioxide or water vapour, that cause a warming of the earth's atmosphere by absorption of infra-red radiation. The term is an inappropriate analogy to green-houses, which were once thought to keep themselves warm by admitting sunlight but retaining infra-red radiation. However, it has been shown that most of the heat retention in green-houses results from the conservation of warm air. The atmosphere, of course, is open, so the mechanism is not the same.

gross national product (*GNP*) The total value of all goods and services produced by the economy in a given year.

half-life (*of a radioactive substance*) The time required for half of a sample of radioactive matter to decompose.

hard energy system An energy system based on large centralised electric-generating facilities. See also soft energy system.

hazardous waste The material that poses a threat to human health and the environment.

heat A form of energy. Every object contains heat energy in an amount that depends on its mass, its temperature, and the specific heat of the materials of which it consists. Heat can also be generated by chemical reactions such as combustion.

heat engine A mechanical device that converts heat to work.

heavy oil Natural petroleum deposits that are too thick to be extracted by conventional pumping techniques. Heavy oils are found in 'dry' conventional wells and also in oil shales and tarsands.

hectare A metric measure of surface area. One hectare is equal to 10,000 square metres or 2.47 acres.

herbicide A chemical used to control unwanted plants.

herbivore An animal that eats only plant matter.

heterotroph An organism that obtains its energy by consuming the tissue of other organisms. See also autotroph.

home range The area in which an animal generally travels and gathers its foods.

hormones These compounds can be used as insecticides, for if sprayed properly they will disrupt the life cycle of specific insect larvae and kill them.

humus The complex mixture of decayed organic matter that is an integral part of healthy soil.

hydroelectric power Power derived from the energy of falling water.

hydrological cycle (*water cycle*) The cycling of water, in all its forms, on the earth.

injunction A legal order requiring a defendant to stop doing a wrongful act.

innate capacity for increase The maximum rate of growth of a population as determined by the reproductive capacity of the organism.

inversion A meteorological condition in which the lower layers of air are cooler than those at higher altitudes. This cool air remains relatively stagnant and causes a concentration of air pollutants and unhealthy conditions in congested urban regions.

isotopes Atoms of the same element that have different mass numbers.

joule A fundamental unit of energy, 4.184 joules = 1 calorie.

kinetic energy Energy of motion.

krill Small, shrimp-like crustacean that grow in large numbers in the cold waters of the southern oceans. Krill represent a primary food supply for many species of whales.

legume Any plant of the family Leguminosae, such as peas, beans, or alfalfa. Bacteria living on the roots of legumes change atmospheric nitrogen, N_2, to nitrogen-containing salts that can be readily assimilated by most plants.

mass number The sum of the number of protons and neutrons in an atomic nucleus.

megalopolis An agglomeration of several smaller cities and towns into a single expanded urban area.

meteorology The science of the earth's atmosphere.

mine spoil The earth and rock removed from a mine and discarded because the mineral or fuel content is too low to warrant extraction.

mineral reserves The estimated supply of ore in the ground.

molecule The fundamental particle that characterises a compound. It consists of a group of atoms held together by chemical bonds.

mutagen A substance that causes mutations. An inheritable change in the genetic material of an individual.

mutualism An interaction beneficial and necessary to both interacting species.

natural selection A series of events occurring in natural ecosystems that eliminate some members of a population and spares those individuals endowed with certain characteristics that are favourable for reproduction.

neutron A fundamental particle of the atom, that is, electrically neutral.

nodules Small pieces of rock, typically having a mass of about 1 kg, that are found on the floor of the ocean. These nodules are rich in manganese and also contain several other minerals such as copper, iron, aluminum and cobalt.

noise Unwanted sound.

nuclear reactor A device that utilises nuclear reactions to produce useful energy.

omnivore An organism that eats both plant and animal tissue. Common omnivores include bears, pigs, rats, chickens and people.

open dump A site where solid waste is deposited on a land surface with little or no treatment.

ore A rock mixture that contains enough valuable minerals to be mined profitably with currently available technology.

organic farming A set of farming practices that uses no synthetic fertilisers or chemicals.

organochlorides A class of organic chemicals that contain chlorine bonded within the molecule. Some organochlorides, such as DDT, are effective pesticides. They are generally broad- spectrum and long-lived in the environment.

organophosphates A class of organic compounds that contain phosphorus and oxygen bonded within the molecule. Some organophosphates, used as pesticides, are broad-spectrum and extremely poisonous, although they are not long-lasting in the environment.

Ozone Triatomic oxygen, O_3.

parasitism A special case of predation in which the predator is much smaller than the victim and obtains its nourishment by consuming the tissue or food supply of a larger living organism.

photosynthesis The process by which chlorophyll bearing plants use energy from the sun to convert carbon dioxide and water to sugars.

phytoplankton Any microscopic, or nearly microscopic, free-floating autotrophic plant in a body of water. There are a great many different species that exist in a community of phytoplankton; these plants occur in large numbers and account for most of the primary production in deep bodies of water.

plankton Any small, free-floating organism living in a body of water.

pollution The impairment of the quality of some portion of the environment by the addition of harmful impurities.

polychlorinated biphenyl (PCB) A class of organochloride chemicals that are structurally related to DDT. They were used widely in the plastics and electrical industries until they were found to be potent environmental poisons.

population The breeding group to which an organism belongs in practice. A population is generally very much smaller than an entire species, for all the members of a species are seldom in close proximity to each other.

population distribution The composition of a population categorised by several variables, often age and sex.

population ecology The branch of ecology dealing with the size, growth, and distribution of populations of organisms.

power The amount of energy delivered in a given time interval. Power = energy/time.

prairie An extensive area of fairly level, predominantly treeless land. Prairies are characterised by an abundance of various types of grasses.

predation An interaction in which some individuals eat others.

predator An animal that attacks, kills and eats other animals.

primary consumer An animal that eats plants.

radioactivity The emission of radiation by atomic nuclei.

radioisotope A radioactive isotope.

radiolysis A chemical change initiated by radiation.

rate of natural increase The difference between the crude birth and crude death rates. Also called the crude reproductive rate.

rate of population growth The increase in population over a given time divided by the initial population.

recycling The process whereby waste materials are re-used for the manufacture of new materials and goods.

redundancy Superabundance. In the context of safety systems, redundancy refers to the provision of a series of devices that duplicate each other's functions and that are programmed to go into operation in sequence if a preceding device in the series fails.

rendering The cooking of animal wastes such as fat, bones, feathers, and blood to yield tallow (used in the manufacture of soap) and high-protein animal feed.

replacement level The level of the total fertility rate that, if continued unchanged for at least a generation, would result in an eventual population growth of zero.

resource Any source of wealth or means of producing wealth.

respiration The process by which plants and animals release energy by the step-wise oxidation of organic molecules.

saltwater intrusion The movement of salt water from the ocean to terrestrial groundwater supplies that occurs when the water table in coastal areas is reduced.

sanitary landfill A site where solid waste is deposited on a land surface, compacted, and covered with dirt to reduce odours and prevent disease and fire.

savannah A type of tropical or sub-tropical prairie that is subject to seasonal patterns of rainfall. Savannahs are common in central Africa.

secondary consumer A predator that eats an animal that eats plants.

selective cutting The process of cutting selected trees in a stand of timber. See clear-cutting.

smog Smoky fog. The word is used loosely to describe visible air pollution.

soft energy system An energy system based on the use of small, home-sized, decentralised devices using renewable resources.

solar cell A semi-conductor device that converts sunlight directly into electrical energy.

solar collector A device designed to concentrate solar energy for a useful purpose.

solar design active systems: Heating systems that use a controlled flow of some substance; usually air or water, to collect, store, and transmit heat from the sun. *passive systems*: Construction systems that use the structure of a building itself to collect, store, and transmit heat from the sun.

solar energy Energy derived from the sun.

species A group of organisms that interbreed with other members of the group, but not with individuals outside the group.

strip mining Any mining operation that operates by removing the surface layers of soil and rock, thereby exposing the deposits of ore to be removed.

subsidence The settling of the surface of the ground as deep ground water or minerals are removed.

subsidies Economic incentives to induce polluters not to despoil the environment.

succession The sequence of changes through which an ecosystem passes during the course of time. *Primary succession* is a sequence that occurs when the terrain is initially lifeless, or almost so. *Secondary succession* is the series of community changes that takes place in disturbed areas previously occupied by a biological community synfuels An abbreviation for synthetic fuels, liquid fuels produced from oil shales, tar sands, or coal.

sustainable development The development that meets the needs of the present without compromising the ability of future generations to meet to meet their own needs.

synthetic chemicals Compounds produced by chemical reactions performed in a laboratory, as opposed to those formed naturally.

technological fix An approach to solving a social problem, such as environmental degradation, by technical or engineering methods.

thermal pollution A change in the quality of an environment (usually an aquatic environment) caused by raising its temperature.

thermodynamics The science concerned with heat and work and the relationships between them. *First Law of Thermodynamics*: Energy cannot be created or destroyed. *Second Law of Thermodynamics*: It is impossible to derive mechanical work from any portion of matter by cooling it below the temperature of the coldest surrounding object.

threshold level The minimum dose of a toxic substance that causes harmful effects.

tidal energy Energy derived from the movement of the tides.

total fertility rate (*TFR*) The total number of infants a woman can be expected to bear during the course of her life if birth rates remain constant for at least one generation.

toxic substance Any substance whose physiological action is harmful to health.

trespass An illegal intrusion or invasion of other people's property.

tropical rain forest A biome that occurs when high, fairly constant rainfall and temperature permit plants to grew rapidly throughout the year.

tumour An abnormal growth of unregulated cells.

tundra Arctic or mountainous areas that are too cold to support trees and are characterised by low mosses and grasses.

urban sprawl The growth of suburban areas and the expansion of cities to pre-empt increasingly large areas of land.

urbanisation A demographic process characterised by movement of people from rural to urban settlements, from small towns to large cities, and from large cities to their suburbs.

vital event In demography, a birth, a death, a marriage, a termination of marriage, or a migration.

water pollution The deterioration of the quality of water that results from the addition of impurities.

water table The upper level of water in the zone of saturated subsurface soil and rock.

zero population growth A condition in which the birth and death rates in a population are equal therefore, the size of the population remains constant.

Bibliography

Arvill, R. (1967) *Man and Environment: Crisis and the Strategy of Choice*, Penguin, Harmondsworth.

Allen, J.P. (1991) *Biosphere 2: The Human Experiment*, Penguin, New York.

Bach, Wilfried (1972) *Atmospheric Pollution*, McGraw Hill, New York.

Bahuguna, S. (1990) 'Tehri: The Dam of Discontent', *Yojana*, Vol. 34 (10) 9–12.

Barrow, C.J. (1995) *Developing the Environment: Problems and Management*, Longman, Harlow.

Brij Gopal and N. Bhardwaj (1979) *Elements of Ecology*, Vikas Publications, New Delhi.

Billihgs, W.D. (1971) *Plant, Man and the Ecosystem*, Macmillan.

Canter, L.W.C. (1996) *Environmental Impact Assessment*, McGraw-Hill, New York.

Clapham, W.B. (1973) *Natural Ecosystems*, Macmillan, London.

Colinvaux, Paul A. (1973) *Introduction to Ecology*, John Wiley and Sons, New York.

Darlingten, P.J. (1957) *Zoogeography*, John Wiley, New York.

Dasseman, R.F. (1976) *Environmental Conservation*, John Wiley, New York.

Detwyler, T.R. (1971) *Man's Impact on Environment*, McGraw-Hill, New York.

Elliott, J.A. (2006) *An Introduction to Sustainable Development*, Routledge, London.

Erocal, D. (ed.) (1991) *Environmental Management in Developing Countries*, Development Centre, OECD, Paris.

Farmer, A. (1997) *Managing Environmental Pollution*, Routledge, London.

Fitzpatrick, T. (2011) *Understanding the Environment and Social Policy*, SPA.

Flanagan, D. (1970) *The Biosphere*, W.H. Freeman and Co., San Francisco.

Furley, P.A. and Newey, W.W. (1983) *Man and the Biosphere*, Butterworths, London.

Futehally, I. (1988) *Our Environment*, N.B.T., New Delhi.

Godrej, D. (2002) *The No-Nonsense Guide to Climate Change*, New Internationalist Publications Ltd., Oxford, UK.

Goudie, A.S (1993) *The Human Impact on Environment* Oxford, Blackwell.

Hunt, D. and Johnson, C. (1995) *Environmental Management Systems*, McGraw Hill, London.

Irving, R.M. and Priddle, G.B. (1971) *Crisis: Readings in Environmental Issues and Strategies*, Macmillan, London.

Jackas, G.V. and White, R.O.(1939) *The Rape of the Earth*, Faber, London.

Kneese, A.V. (1977) *Economics and the Environment*, Penguin, Harmondsworth.

Maddox, J. (1972) *The Doomsday Syndrome*, McGraw-Hill, New York.

Manivasakam, N. (1992) *Environmental Pollution*, N.B.T., New Delhi.

Maxwell, K.E. (1973) *Environment of Life*, Dickenson Publishers, Co., California.

McCormick, J.F. (1992) *The Global Environment Movement*, Belhaven, London.

McGill, S. (1999) *Environmental Risk Management*, Routledge, London.

Nijkamp, P. (1980) *Environmental Policy Analysis*, Wiley, New York.

Odum, Eugene P. (1971) *Fundamentals of Ecology*, W.B. Saunders Co., London.

Osborn, F. and Bigg, T. (1998) *Earth Summit II: Outcomes and Analysis*, Earthscan, London.

Owen, L. and Unwin, T. (eds.) (1997) *Environmental Management*, Blackwell Pub., Oxford.

Park, C.C. (1980) *Ecology and Environmental Management*, Butterworths, London.

Munasinghe, M. (2009) *Sustainable Development in Practice*, Cambridge University Press, Cambridge.

Pfafflin, J.R. and Ziegler, E.N. (eds.) (1976) *Encyclopedia of Environmental Science and Engineering*, Vols. I and II Gordon and Breach Science Publishers, New York.

Redclift, Michael (1987) *Sustainable Development*, Methuen, London.

Saxena, H.M. (2004) *Environmental Geography,* Rawat Publications, Jaipur.

Saxena, H.M. (2010) *Environmental Management,* Rawat Publications, Jaipur.

Sandgrove, K. (1992) *The Green Manager's Handbook*, Gower, London.

Samson, F.B. and Knopf, F.L. (eds.) (1996) *Ecosystem Management: Selected Readings*, Springer, New York.

Saraf, R.K. (eds.) (1973) *Proceedings of Symposium on Environmental Pollution*, CEPHERI, Nagpur.

Seshadri, B. (1969) *The Twilight of India's Wild Life*, Oxford, London.

Sewell, G.H. (1975) *Environmental Quality Management*, Prentice-Hall, New York.

Stephen, Tinsley (1999) *Environmental Management Plans Demystified,* Spon Press, London.

Sharma, H.S. (eds.) (1988) *Environmental Design and Development*, Scientific Publications, Jodhpur.

Strahler, A.N. and Strahler, A.H. (1976) *Geography and Man's Environment*, John Wiley, New York.

Thomas, W.L. (1965) *Man's Role in Changing the Face of the Earth*, University of Chicago Press, Chicago.

Turk, J. (1985) *Introduction to Environmental Studies*, C.B.S. College Publishers, Chicago.

UNDP (1992) *Handbook and Guidelines for Environmental Management and Sustainable Development*, New York.

UNESCO (1981) *M.B.A. Information System Biosphere Reserves*, Publishers, No. 2.

UNESCO (1970) *Use and Conservation of Biosphere*, Natural Resources Research 10, Paris.

UNO (1992) *Agenda 21: Programme of Action for Sustainable Development,* UN Publication.

Watt, K.E.F. (1969) *Ecology and Resource Management*, McGraw Hill, New York.

Whitehead, M. (2007) *Spaces of Sustainability,* Routledge, London.

Zimmermann, E.W. (1951) *World Resources and Industries*, Harper and Row, London.

World Commission on Environment and Development (1987) *Our Common Future,* Oxford University Press. New York.